SpringerBriefs in Energy

More information about this series at http://www.springer.com/series/8903

David Bonilla

Air Power and Freight

The View from the European Union and
China

 Springer

David Bonilla
National Autonomous University of Mexico,
(Universidad Nacional Autónoma de México)
Institute of Economic Research
Mexico, Mexico

Transport Studies Unit,
University of Oxford,
Oxford, UK

ISSN 2191-5520   ISSN 2191-5539   (electronic)
SpringerBriefs in Energy
ISBN 978-3-030-27782-6   ISBN 978-3-030-27783-3   (eBook)
https://doi.org/10.1007/978-3-030-27783-3

This Springer imprint is published by the registered company Springer Nature Switzerland AG
The registered company address is: Gewerbestrasse 11, 6330 Cham, Switzerland

Foreword

The increasing globalisation of the world economy means that transport becomes the key element in ensuring connectivity and the means by which the complex and lengthy supply chains can be maintained under all conditions. Small delays or a lack of resilience anywhere in these chains will result in disruption, increased costs and the need to re-evaluate the risks to trade and movement. But it is not just the physical capacity of the global and the local transport systems to respond to resilience, as there are also equally important concerns over governance and responsibilities, the levels of technological dependence, the flows of information and the monitoring of individual transactions, and even the stability of the different international regimes. This is an important agenda that is addressed in David Bonilla's new book, as the outcomes of decisions made by governments, businesses and individuals will determine the future direction of global trade.

The necessary conditions for continued economic prosperity are that strong global governance is required so that scale economies in mega infrastructure project investments (such as China's Belt and Road Initiative) can be achieved, and at the same time the appropriate regulation of air transport and freight movement can be implemented to minimise the environmental impact. Even though there may be signs that freight dematerialisation is taking place, this is likely to be more than countered by the continuing offshoring of manufacturing, which in turn is likely to increase the transport content, as the distance between the locations of production and consumption of goods increases. Indeed, given the current patterns of carbon dioxide production, it is likely that transport will increase its contribution from the current 25% (2019) to over 50% (2050), as other sectors reduce their carbon dependence, and as air and maritime transport grow in importance. At present, both international air and maritime transport are not subject to any of the global climate agreements, and these two key freight transport modes have only recently engaged in the climate change debates.

This book on 'Air power and freight' adds value to the complex and rapidly changing picture of the movement of goods through a series of national case studies that relate freight transport levels to a range of economic and social factors. This analysis suggests that convergence is taking place between the national patterns of economic activity and the quantity of transport needed. It is also concluded that there is some evidence of improved efficiency, as unit transport costs have been reduced. It is important that lessons concerning best national practice are transferred between the different countries so that substantial rather than incremental change can take place towards sustainability.

Road freight is currently dominant at the national level, but airfreight becomes more central when high value to weight products are concerned, or where time is of the essence. The global patterns of airfreight have developed through hub-and-spoke operations, where shorter direct movements are replaced by longer more indirect movements. This means that distances are increased, with additional environmental costs, but there are cost savings from network economies. As with many of the decisions in the transport sector, there are difficult choices to be made about how to minimise costs, but at the same time to address the broader concerns over planetary boundaries.

David Bonilla has also placed considerable effort in bringing together a wide range of empirical evidence to augment the important policy and global governance issues raised in his book. Included here is material from the EU (including the UK), Japan, China, Mexico and the USA. The selection reflects the availability of high quality data, but it also covers the major global trading nations and encompasses countries at different stages of development. As such the aim is to represent a range of different situations, but it is also important to include the major players so that each can learn from the others. Such learning must provide one of the main mechanisms through which global consensus can be achieved, and this necessitates the engagement of politicians, business leaders, and multi-national companies in high level debates that lead to positive outcomes.

The timing of the book is also opportune, as fundamental issues concerning the health of the planet and the future of globalisation in a technological and knowledge-based society are both key discussion topics and urgent action is needed now. Global freight patterns should provide a substantial element in those discussions over whether global society can continue to promote unsustainable transport. The recent geo-political moves towards populism and trade barriers provide a second set of new constraints that may also question the more conventional approach of reducing global trade barriers. In the short term, protectionism may help to dampen the growth rate in freight transport, but in the longer term there will also be considerable redirection of trade flows as new patterns of trade emerge. Eventually, those recently raised barriers may again be reduced. It is important that such a 'breathing space' should be productively used to substantially reduce the environmental and social impacts of long distance freight transport.

David Banister
Professor Emeritus
Transport Studies, School of Geography
and the Environment
University of Oxford
Oxford, UK
May 2019

Endorsement

This book is a welcome addition to the relatively sparse literature on the economic geography of air cargo and road freight trends. By assembling a wealth of statistical data and consulting a range of stakeholders, Bonilla has analyzed these trends from trade, locational, and environmental perspectives and considered where they may lead. This makes the book particularly topical in a world in which protectionism, geopolitics, and climate change are increasingly shaping public policy and business strategies on freight transport. While it focuses on goods movements in Europe and China, its conclusions also have wider relevance at a global level.

Alan McKinnon
Professor of logistics
Khune Logistics University

Acknowledgements

To my parents Gabriela and Arturo for giving me the courage to write.

To my daughter Mia and to my wife Hao Feng for having put up with my late nights.

The author is grateful to the Institute of Economic Research (Instituto de Investigaciones Económicas, Universidad Nacional Autónoma de México). Also special thanks to the Transport Studies Unit, Oxford, in particular for his encouragement to write this book, Prof. David Banister, former Director (2007–2015) of the Transport Studies Unit, Oxford. David was an excellent mentor while I was a Research Fellow at Oxford.

Last but not least I thank Prof A. McKinnon for his comments on an earlier draft.

This book has been written alongside seminars where I received comments. The seminars were held in many locations in UNAM, Mexico, and previously in Oxford, Brussels, in London, in Vienna and in Hong Kong. I started to sketch out the structure of this book in Oxford and ended up writing it in Mexico city.

Contents

Chapter 1
Introduction

Without transport there can be no globalization, neither connectivity of people nor ideas, nor regional trade. This book provides a critical assessment of surface freight transport and of airfreight movements in key continents for the last decades of the twentieth century and for the present one. By 'critical assessment', one means doing a careful evaluation of how the relationships among freight transport activity, trade and space interact with each other. China is registering higher economic growth, vigorous export performance and a growing air cargo market both domestically and offshore. In contrast to China, the European Union and the North America regions (in this book, mainly the USA and Mexico) are registering far lower economic and trade expansion and within these two regions there are winds that favour the adoption of protectionist trade policies under the banners of Brexit and Trump's 'America First'. It remains to be seen how global trade policy is bent towards protectionism, which will inevitably lower the expansion of all transport modes: air, road, ocean, rail and waterways and of the entire freight transport sector. The recent inclusion of Italy within the Road and Belt initiative (R&B) which will extend from East Asia, mainly China to Europe is likely to stimulate the growth of global and European freight transport. The intent of R&B initiative involves Chinese investment in a network of infrastructure projects connecting Asia, the Middle East, Africa and Europe.

The approach in the book is largely qualitative but publicly available data is used as much as possible to understand the historical changes in the air transport and the road sectors using an economic geography perspective and the tools of transport economics. The geographical regions selected for analysis offer many advantages: the European Union, the USA and Japan are high income regions, which provide many policy lessons which China or others can learn from. Air to rail integration can be one possible solution in the medium term.

Chapter 2 lays out the ground for the rest of the book by highlighting the rapid historical growth of road freight and of air transport after World War II in key West European, Asian and Latin American economies.

The *key* questions addressed are described in the following chapters of the book:

© Springer Nature Switzerland AG 2020
D. Bonilla, *Air Power and Freight*, SpringerBriefs in Energy,
https://doi.org/10.1007/978-3-030-27783-3_1

In Chap. 2, we focus on the road freight industry. The road freight transport sector has registered rapid historical growth in key Western economies, and more so in emerging economies, this growth requires diesel fuel to power the sector. There is a need to control the growth of the sector to reduce fuel use and this leads to two questions. First, how much of this growth is explained by the rebound effect (RE)? Second, will the annual growth rate of diesel use of road freight transport slow or even turn negative after the adoption of energy efficiency measures?. The chapter presents a cross country overview of key driving forces of road freight: competition among the modes, decadal changes in distance travelled by trucks for various countries, i.e. the USA, China and others which have contributed to road congestion.

By means of a simple econometric model we explain the traffic of road freight flows, we present an overview of how energy use of road freight transport may increase under the effect of the rebounding energy use. We find common challenges, among advanced economies with similar post-industrial features and common trends of personal income levels, of average truck haul, of truck loads and of population density levels.

Chapter 3 expands the coverage of road freight transport by country. The key questions of the Chapter are: How desirable is it to reduce the volume of freight transport in the European economies? How feasible is this reduction in 2050? What is the role of stakeholders in setting the energy efficiency targets of road freight transport sector in the European Union and the policy agenda? The chapter presents an overview of key environmental limits of road freight flows (energy use and Greenhouse gas emissions) and that sectors' contribution to global warming and other sustainability considerations which require urgent action.

In Chap. 3, we propose an image of the future or a vision to achieve a sustainable road freight transport system; and provide transport policy solutions for the sector to 2050. Following the step by step description of the practice of backcasting, the method is applied to idealized pathways of freight transport and of diesel use to formulate measures to strengthen their acceptance. Our findings confirm that by using the knowledge of stakeholders, we can propose a framework to identify policy targets, build story lines, and find key megatrends which shape the growth of the freight sector to build the vision. Through the insights of stakeholders we propose a vision which allows transport policy makers and corporations involved in freight transport in the European Union to take actions to face threats effectively. The vision is a reaction to the megatrends.

In Chap. 4, we discuss the role and status of airfreight transport in the European Union and in Western Europe in particular. The overall goal of this chapter is to determine what role the airfreight sector plays in leading to regional growth in Europe and in particular in North-West Europe? Airfreight transport has grown on an annual average of 6% in 1980–2010 in Europe. The sector, however, faces limits for expansion in coming decades. In the chapter we identify key sources of growth of airfreight in the European Union and (a) assess how the geography of the airfreight sector determines the location of logistics firms and (b) scan patterns in future threats of air cargo firms (i.e. the limits to growth of airfreight volume) and opportunities for them. This scanning exercise guides the sector on potential courses of action. We identify key developments of the airfreight transport sector based on (a) airfreight trends, including CO_2 emissions and local emissions and (b) related exports and imports of 1970–2010 in Europe. We also observe that the location of logistics firms can be explained by clusters of airfreight activity across North-West Europe. Finally, we disaggregate airfreight movements of European Union 27 nations to explain its growth. The evidence shows the airfreight sector increasingly integrates supply chains, through expansion strategies of transport firms, enabling air cargo flows and cities to grow. Airfreight incubates economic activities and shapes the logistics firms' spatial behaviour. The evidence furnished shows that the airfreight mode raises profit levels of logistics' firms. The cumulative result of hub–spoke networks, in North-West Europe, is to increase the supply chains' connectivity and to alter the geography of airfreight through broader hubs.

The key questions of Chap. 5 are the following: What are the sources of growth of airfreight in China in recent decades? How has growth of regional freight transport been supported by the development of airports in China's regions? And how have the economies of Chinese cities benefited from the growth in the number of airports, airport capacity and freight flows?

In Chap. 5, we unravel the economic and geographical forces that have led to the rapid growth of the airfreight sector of 285 cities in China. Air transport is a fast growing sector. A recent factor for growth in airfreight is the spread of e-commerce, that of express mail in warehouses, in trucks and in entire supply chains (Kasarda and Lindsay 2011). Many other drivers of growth of the air transport market are found. Three additional drivers of the airfreight market stand out. One is the adoption of hub and spoke networks which have increased both freight volumes and connectivity. The second driver is the various airfreight/network carriers that serve the local market and are connected to other airports at the same time. Due to the consolidation of the European Union and US air transport industry, more and more network carriers operate out of multiple connecting hubs (Burghouwt 2013). A third driver is the adoption of the container in supply chains. Using statistical material on airports, on freight volume, and on socio-economic features, the evidence shows the influence factors of airfreight growth can be spatial rather than only economic and sometimes geography is a more powerful explanatory variable.

In Chap. 6 we discuss the conclusions on two important modes of freight transport for global trade (Air transport and road transport). We study the entire modern history of the two sectors which leads to key solutions to achieve a more sustainable freight transport system and on how the sector can contribute to *global* warming mitigation.

This monograph is distinctive for three reasons. First, there are few comparative studies on freight transport developments for strategic nations such as the USA, China, Japan, the UK and the European Union; second there are few studies with an applied economics focus on the freight transport and global economy relationship. Third, there are hardly any studies on China's airfreight developments that cover both periods of rapid and slow economic growth with a lens of economic geography;

neither there are studies that observe air transport flows at the regional levels as we do in this volume.

In the case of the backcasting freight transport in the European Union, the book is distinctive in three ways. First few studies have been performed on freight themes within a future perspective. Second, few studies rely on collaborative and participative methods to collect data and formulate some of the objectives and hypothesis. Exceptions in the literature are works by Macharis et al. (2010). Third, this chapter is in line with the works of Van Duin (2012) and of Gonzales Feliu et al. (2013). Backcasting can support decision making within the freight transport sector based on theories on group reasoning as described in Raifa et al. (2002).

Therefore the multi-stakeholder nature of backcasting can support decision making. This volume adds to that literature.

As far as studies on airfreight transport, the volume follows the tradition in the freight and logistics literature of Levinson's 'The Box', Kasarda and Lindsay's (2011) Aerotropolis; and Gilbert and Perl's (2008), 'The Transport Revolution') all of whom have examined freight transport flows using case studies and empirical data. As well as using evidence from case studies we use quantitative models and data to support our findings. Unlike these authors we use recent evidence from key developments in world regions that are shaping freight transport activity at this moment, the relationships across different freight transport markets and their transport and the environment nexus. The airfreight literature is lacking a holistic approach at the sectoral and historical levels; often much of the literature assesses air transport outside the realm of developments of the global and regional economies.

References

Gilbert R, Perl A (2008) Transport revolutions: moving people and freight without oil. Earthscan, London

Gonzales Feliu J, Morana J, Salanova Grau J, Ma TY (2013) Design and scenario assessment for collaborative logistics and freight transport systems. Int J Transp Econ 40:207–240

Kasarda JD, Lindsay G (2011) Aerotropolis: the way we live next. Penguin, London

Levinson M (2006) The box: how the shipping container made the world smaller and the world economy bigger. Princeton University Press, Oxfordshire

Macharis C, De Witte A, Turcksin L (2010) The multi-actor multi-criteria analysis (MAMCA) application in the Flemish long-term decision making process on mobility and logistics. Transp Policy 17(5):303–311

Raiffa H, Richardson J, Metcalfe D (2002) Negotiation analysis: the science and art of collaborative decision making. Harvard University Press, London

Van Duin R (2012) Logistics concept development in multi-actor environments: aligning stakeholders for successful development of public/private logistics systems by increased awareness of multi-actor objectives and perceptions. PhD thesis. TRAIL thesis series T2012/6, The Netherlands TRAIL Research School

Chapter 2
Road Freight Transport and Energy Use: The USA, China, the EU, Japan and Germany

We find that RE estimates of energy use of moving cargo (Megajoules/t-km) differ strongly from country to country. We observe a upward trend in truck freight fuel intensity for (energy use per tonne-km moved), and on-road truck fuel economy for the smaller nations (japan, Denmark). A downward trend in fuel economy is found for the large economies (China, the U.S.). We also observe falling diesel prices. Despite this truck km travelled continues an upward trend in many cases. This chapter contributes to understanding the primary definition of the RE.

2.1 Introduction

There are many studies on the rebound effect (RE) for the passenger sector. See Sorrell and Dimitropoulos (2008) and Sorrell (2007) for a review. But there is a clear gap in the literature of energy economics and of transport studies. The RE is important when assessing the net effects of energy efficiency measures (i.e. engine fuel efficiency, fuel switching, telematics, etc.). Two studies estimate demand elasticities of road freight which can be used to calculate the RE in some way. First is the work of Barker and Kohler (2000) who develop an analysis for charging for road freight activity using the macroeconometric model of the EU. Bjorner (1999) examines the Danish freight transport sector by applying the vector autoregressive model to work out elasticities of the sector. A study of the 4CMR (2006) research institute calculates the direct and indirect effects of the RE on the UK economy using an extensive data set for road freight and other modes. Addressing an important policy question the EC has funded a project on the RE (EC 2012).

© Springer Nature Switzerland AG 2020
D. Bonilla, *Air Power and Freight*, SpringerBriefs in Energy,
https://doi.org/10.1007/978-3-030-27783-3_2

This chapter seeks to answer two questions. How can one measure how large or small the RE is? To answer the questions we use data on trade activity, population clusters in cities, the distance travelled by trucks (truck-km), and on inter-modal competition. Second, is there a common explanation, of the RE, across trade intensive economies with similar post-industrial features, or population density levels?

The rebound effect is defined as the difference between projected energy savings and actual ones resulting from enhanced energy efficiency (Khazzoom 1980; Sorrell 2008; and others). This is also known as the Brookes' postulate (Brookes 1990).

We estimate the rebound effect for the 1978–2008 period for three countries using evidence from quantitative history and econometric models. Using data on freight transport services, we estimate the rebound effect for road freight transport (t-km; a truck load of moving a 100 km counts as 100 t-km, IEA 1997; Schipper et al. 1992) of Denmark, Japan and the UK and assess the implications of falling energy prices on road freight activity and on total energy use (mainly diesel fuels). To estimate the effect of energy efficiency improvements, the historical changes in the RE for freight transport growth are estimated using microeconomic evidence.

For example, Japan's road freight sector has seen an improvement of 49% (kilocalorie/t-km moved) of its unit energy efficiency in 1965–2008 $[1429 - 722/1429 * 100 = 49\%]$ (IEEJ 2010) but if this produced a 30% reduction in total energy use (of freight moved by road) the rebound effect would be $[49 - 30/49 * 100 = 39\%]$. Energy use in road freight activity should have fallen by the same amount of the unit improvement (in kilocalorie/t-km) but in fact energy use of that sector has increased 150% in that period. The 10% $[49 - 39]$ of energy savings that is missing, from the expected savings, represents the extra energy use by the new more efficient truck i.e. because the freight forwarder adopts more efficient engines allowing to move freight for even longer distances. This means engineering based estimates of energy savings can be overestimated if energy efficient policies, or sustainable transport ones, ignore the size of the rebound effect for the road freight sector.

Regarding the rebound effect, the EPA (2011) of the U.S. has concluded that 'the fuel economy rebound effect for light duty vehicles has been the subject of a large number of studies since the early 1980s. Although they have reported a wider range of estimates of its exact magnitude, these studies generally conclude that a significant rebound effect occurs when fuel efficiency improves'. EPA (2011) quoting the EPA and NHTSA (p. 44, EPA).

Rebound effect (RE) theory has been criticized extensively in debates on resource management. Leading criticisms of RE centre on 'many of the hypothesis of recent papers promoting RE are neither scientific nor testable' (Goldstein et al. 2011); and on "too much focus on correlation" (between energy efficiency improvements and extra energy use); the literature also focuses on "too much causation" (Nadel 2012), and "on the size of the RE" (Schipper and Grubb 2000). Nadel argues that the chain of causation in explanations for the effect are unclear.

Further criticisms of RE theory include 'the assumption that the RE is consistent and universal across uses and levels of efficiency (Goldstein et al. 2011)' and that 'the evidence for the RE is weak' (ACEEE 2012). Jevons (1865) first proposed the concept of RE through his analysis of the coal use in British industry. In a critique on RE published in the New Yorker (Owen 2010), it was argued that '…the young, dynamic industrial world, Jevons lived in, no longer exists'. A second critique (Koomey 2011) emphasizes the lack of inclusion of income effects in the studies of RE and cautions against the argument that 'the energy efficiency savings are offset 100%'.

However, there is a need to explain (a) the growth of energy use (or other natural resources in general) in end use sectors and (b) the absence of correlation between that energy use and increasing energy efficiency gains. The rebound effect can explain the growth of freight energy and increasing energy consumption of freight transport.

Freight transport demand is defined as the product of goods lifted times hauled distance and is measured in t-km. Road freight transport in Japan, has grown by 1.0% per year in 1980–2010 below the growth of the UK and that of Denmark in the same period (1.5%: for 1980–2006) (Based on data in: Stat Bank Denmark 2008, 2011; IIEJ 2010; TSGB 2013). Past trends (1973–1992) indicate that the road freight mode in Denmark expanded by 54% (Schipper et al. 1997) and by 48% in 1990–2005. The growth of road freight transport, however, in the three economies can reflect different logistical patterns that are embedded in each nation's geographical features (levels of dispersion of warehousing sites, of production and of consumption centres).

This analysis can guide policymakers on whether, and on how, the RE reinforces the long-term trend between road freight transport (t-km, Fig. 2.1), energy efficiency gains and the economic growth. To answer these questions, data on (1) national economic activity and (2) on freight transport movements, for world leading economies, is used.

2.2   Background of Key Drivers of Freight Transport

There are five key drivers that we discuss in this section: the volume of trade, (also discussed in Chap. 5) population density (population in cities with more than one million), the average length of haul of trucks, how far these travel and the degree of competition among the transport modes. A key driver of freight transport is trade. The latter accounts for 41% of China's GDP and for 28% in the case of the USA and 85% of the Euro Area (World Bank 2018). The Euro area is far more trade intensive (trade as a fraction of GDP) than the other economies. Trade accounts for 30% of Japan's GDP; for 100% of Denmark's and 87% for Germany and 78% of Mexico's GDP in 2016 (Ibid.). However, the modes used to move freight of each of the three regions will differ substantially. The role of diesel prices is discussed at the end of the chapter.

Though geographically apart, the USA and China share similar geography in terms of size (the USA: 9.6 million square kilometre and China with 9.5 million square kilometre; World Bank 2009) and soon to be identical t-km per capita. These two however share extremely different levels of population density and urbanized

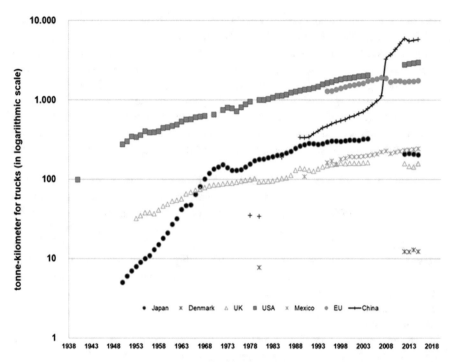

Fig. 2.1 Growth of road freight transport (U.S: Bureau of Transportation Statistics 2018, US; Eurostat 2018; National Bureau of Statistics China (2018); Statistics Denmark 2012, (various years). Japan data in IIEJ (2011, 2017) and the World Bank (2013). Japan data includes vans, HGVs and mini vans. UK data (DfT 2011; DECC 2016) includes vans and HGVs; Danish data includes vans and HGVs. Mexico, OECD (2018)

transport infrastructure with the USA taking the lead over China. Despite the higher trade intensity of China, population centres with more than one million account for almost a fifth of China's population. In contrast to China half of the US population live in these classes of cities (based on data in World Bank 2009), reflecting both a higher urbanization level for the USA and a higher spatial concentration of US economic activity.

Within the largest nine economies of the EU less than one-fifth of the population live in cities, on average, of more than one million people (data in World Bank 2009). The spatial concentration of economic activity is therefore much higher in the USA than within the EU. The clustering of population centres in these sorts of cities is higher in Mexico than in the EU States but well below the US level. Japan has a higher fraction of cities in this rank than does China or the UK. Overall, the level of clustering of population is highest in the USA and Japan than anywhere else in our sample. The same applies to the spatial concentration of economic activity. In this regard, China has a slack for further spatial concentration of economic activity in the near future.

Japan and Denmark share similar features: levels of t-km per capita moved by road freight transport, similar hauling distance and population density. These nations are highly trade intensive economies, have limited habitable land and show a highly urbanized transport infrastructure. Population centres with more than one million account for half of Japan's population and for one-fifth of Denmark's population (World Bank Data 2013). This means urban population density is higher 7 in Japan than in Denmark. Population density levels off at 350 people per square kilometre in Japan and 130 in Denmark. The density level reveals that the spatial concentration of Japan is higher than Denmark's.

2.2.1   Understanding Growth Patterns of Freight Transport in Large Economies

The freight industry is an important part of world leading economies but it is not always clearly defined in the literature: data for the sector sometimes includes freight transport for all modes and warehousing segments; in other cases, the warehousing data is not reported in official statistics. The industry represents 6.5% of the EU economy (Eurostat 2017, all modes); 6.8% for China (China federation of logistics and purchasing, 2018; all modes, warehousing), 5.3% for the USA (all modes, Bureau of Transportation Statistics 2018); 4.7% of Japan' GDP, all modes are considered (Data Monitor 2011). The UK freight sector (all modes) accounts for 4% of national GDP (data from the World Bank 2018 and the UK Institute of Mechanical Engineers 2013). The German freight sector accounts for 9% of GDP (using data of Eurostat 2018) but this figure includes transport, trans-shipment and storage of goods and related services such as the forwarding business or the operation of freight centres, ports and other infrastructure facilities (Deutsche Bank 2015).

Freight transport (road, rail, internal navigation and air modes) in Japan accounts 248 billion US$ in 2010 (Data Monitor 2011); whilst road freight alone accounts for 110 billion US$ in 2010 (Data Monitor 2011).

In contrast to Japan, Denmark's freight industry represents 3.4% of national GDP with a value of nearly 40 billion Danish Kroner (DKr) in 2006 (Statistics Denmark 2012, in fixed prices of year 2000). In recent decades, these countries have seen persistent increases in freight transport activity (Fig. 2.1) with the fastest increase recorded in China.

Throughout modern history, freight transport growth is fastest in Japan (1948–1970), and more recently in China (2002–2008). Figure 2.1 shows the growth for six countries or regions (the EU, the USA, China, Japan, the UK, Denmark and Mexico for years 1948–2015). The data is presented in magnitudes rather than in absolute figures in order to show the rate of growth of freight transport across years. A few data points of Fig. 2.1 are much larger than the bulk of the data on road freight transport. The data ranges from 10 to about 10,000 in the logarithmic scale.

The EU and the USA show similar volumes of freight transport demand and growth trends. In 2003, the growth path of China suddenly shoots upwards and overtakes the rate of growth of the EU and of the US freight transport sectors (road only). This uptick in road freight transport reflects the size of the manufacturing sector of China and of exports of that sector. The US rate slows down 1995–2015 as a result of lower manufacturing output.

In the 1950s, Japan shows the fastest growth in freight transport followed by that of the UK and the USA. The slope of data (Fig. 2.1) for Japan is the steepest in that period until 1968. Japan's growth rate overtakes in the 1960s that of the UK while Mexico's growth rate overtakes Japan's in the 2000s as the latter losses industrial capacity; Mexico, however, does not match the fast rates of Japan recorded in the 1950s or 1960s. In the 1980s, Mexico overtakes the UK as far as the road freight moved. The slope for data on China is also steepest in the sample (Fig. 2.1) and the slope of data for the UK is flatter. It is possible to expect a flattening of the China trend should the latter follow the US trend from a global producer to a global consumer of products.

Small changes in China's growth rate of regional or global trade can have ripple effects on other countries such as Mexico, the USA, Japan and the EU. Those changes can magnify a decline in freight transport sectors of those nations. It is entirely feasible to assume that the rate of growth of Mexico or the USA would have been higher in the absence of higher than average growth of China's surface freight within the 2000s. Both China and Mexico are low cost manufacturers that may be competing for similar markets such as that of the US one.

The rapid increase of China's freight transport sector results from industrial flight from the USA, or Japan, to China, or from the EU to the latter. Industrial flight refers to the transfer of industrial activity from one country to another. This is not shown by the data within the decade of 2000s but we can assume the rate of growth of the US freight would have been higher in the absence of China or in the absence of industrial flight from US to China or Mexico to China. In short the large increase in industrial capacity induces the strong growth in freight transport activity in China and abroad.

Denmark shows a steady path in 1978–2015, a path similar to that of the UK road freight sector; this trend reflects mature economies that increasingly rely on services which are less freight intensive (Agnolucci and Bonilla 2009). In 2015, China's road freight transport is 28 times larger than that of Japan and even larger than Denmark's. And it took China just a few years to achieve the level of road freight transport what took decades for the USA or other economies. During 1955–2015, China's road freight sector registers growth of 1000% whilst the USA, the EU, and Mexico register a combined average of 36%; Japan and the UK as a result of slower growth in manufacturing output showed a growth rate of −50% and of 15%, respectively.

Figure 2.2 shows the time path of the road freight transport using (per capita levels: t-km per population) for selected years (1990–2015) for the previously listed countries plus Mexico, this one is an USMCA trade treaty member).

In three country regions (Japan, the USA and the EU) freight transport grows exponentially during the 1950s to the 1970s (Fig. 2.2) and rates of growth decline in all countries from the 1980s to 2015, except for China. Although road freight demand (Fig. 2.1) has a forecastable aspect, its random features make it difficult to accurately predict its future path.

Even today, the US displays the largest freight movement on a per capita basis as it has transitioned from the empire of production (World War II to the 1970s) to the

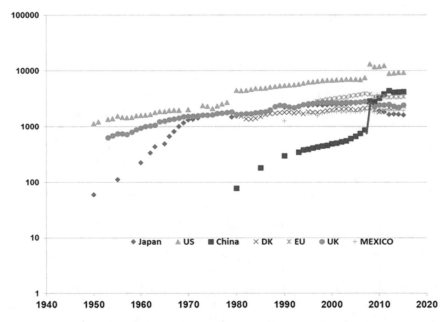

Fig. 2.2 Growth of freight transport movements per capita (road mode). (Sources: see Fig. 2.1. Population data: World Bank development indicators (2018))

empire of consumption (1970s to 2016; Maier 2006). US leadership in production and consumption relies on freight transport. The EU 28 Members States are about a fifth of the US level (tons moved by road). China stands at around 50% of the US level in 2016. However, it took the US economy 58 years (1950–2008) to achieve what China achieved in only 4 years (2007–2011). One can see the speed of China's take-off in terms of freight moved. It took Japan 24 years to double (1980–2004) its per capita level while it took China only 9 years (1985–2004) to match Japan's level. This cross country comparisons illustrate the speed of economic growth of China's economy.

These trends reflect the manufacturing success of China and to a lesser degree of the EU and US economies. What is striking is the speed and breadth of China's take-off in transport activity.

Some economies are more freight intensive than others. The EU average (27 Member States) stands at 238.9 t-km per GDP in 2005 (European Commission 2005), the USA at 181 (2015) peaking in 2011 at 253 t-km per 1000 GDP and China at 768 per 1000 GDP (World Bank 2018). China's road freight intensity is the highest in the sampled countries. This means that to produce 1000 of GDP activity the economy requires inputs of freight transport. Road freight intensity levels do not always correlate to trade intensity ones, e.g. the Euro area is both highly road freight intensive and trade intensive but the EU's freight intensity lies in the same range of an economy with a much lower level of trade intensity e.g. the USA. Other factors explain this intensity such as the stiff competition, or the lack of it, from other modes for moving freight.

It is worth noting that Japan and Denmark are not particularly freight-intensive countries. At 60 t-km per 1000 GDP, Japan's freight intensity level, in 2010, matches that of Denmark (based on data in: World Bank 2013; IIEJ 2011; Statistics Denmark 2012). The UK's ratio is about 80 in 2010 (data in World Bank 2013). These three nations share a common feature: they have smaller territory which limits the volume of national freight. With a far larger territory and plenty of trade volume, Mexico's freight intensity peaks at 227 t-km/1000 GDP in 1999 and stabilizes at 199 t-km/1000 GDP in 2016 (Data calculated based on World Bank 2018)

China's carbon intensity is 40 (kg-CO_2/1000 t-km) (data converted from Hao 2017, p. 8) in 2014, this level is low compared to other countries but this can be explained by the dominance of long distances of road freight in China's territory as observed in Hao et al. (2017). In contrast to China, Japan's shorter distances explain its higher carbon intensity (road freight) of around 213 (kg-CO_2/1000 t-km) in 2015 representing a 20% decline if compared to the 1980 level.

As in China, US carbon emissions are also explained by long distance freight transport which are around 26 kg-CO_2/1000 t-km (using data in Kamakate and Schipper 2009). The EU 28 MS register emissions of 87 kg-CO_2/1000 (2005) (using data of 2005 road only). These trends suggest that the EU road freight transport sector is more carbon intensive than those of the USA or China. The EU's carbon intensity may be explained by shorter distances of the movement of freight within Europe's highways and cities.

Denmark's intensity, of its entire truck freight fleet, is one of the highest in the OECD: it stands at around 188 (kg-CO_2/1000 t-km) in 2005 which represents an increase of 160% since 1980 (IEA 1997). In 2009, the UK ratio is 124 (kg-CO_2/1000 t-km) and it has only declined by a small margin since 1980.

The high carbon (and fuel) intensity of Japanese and Danish trucks mirrors two things: (1) the popularity of smaller trucks and vans (with a lower fuel economy performance) with a higher ratio of Litre of diesel: L/100 km; (2) the growth of the length of haul of freight moved by truck and (3) the increasing frequency of empty trips of van deliveries, a practice usually correlated to van usage.

2.2.2   How Does the Freight Move Among Modes?

Figure 2.3 depicts changes in the modal split in 1973 and 2010 for selected economies. Japan's reliance on the road mode continues to widen in that period, alongside the UK and Denmark. Out of these economies, Japan widens its dependence on the road mode strongly. The market share of rail freight continues to decline in Japan and Denmark, whilst the UK rail share increases (Fig. 2.3).

In recent years, the modal split of freight movements reflects the competitive advantage of each mode.

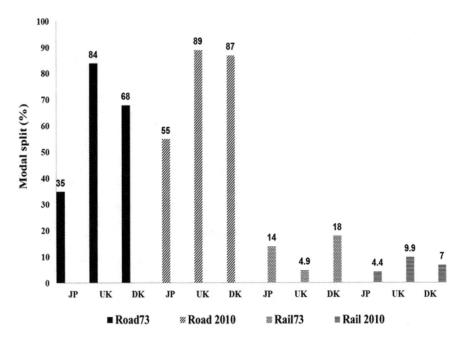

Fig. 2.3 Modal split for freight transport: Japan and Denmark. (Source: Schipper et al. (1997), Japan Statistical Yearbook (2013), European Commission (2006), Stat Bank (2012)). Denmark; the latest data refers to 2005. % share are based on t-km data. These figures exclude freight moved by pipelines. Refer to Fig. 2.1 for extra data sources

The EU freight transport market is far more heavily dominated by the road mode than the economies of the USA, China and Japan; the latter two rely on lower carbon intensive transport modes (Table 2.1) e.g. rail or inland waterways. China shows the lowest dependence on the road mode out of all these economies whilst Denmark shows the highest dependence on that mode because that country's transport system requires shorter distances (or shorter length of haul) (Table 2.1). Germany's transport system also looks vulnerable to the road mode widening its energy imports unnecessarily. Similarly, the UK road freight transport sector reveals high dependence on that mode. These trends are raising concerns among policymakers for further decoupling of road freight from the wider economy (Loo and Banister 2016; Gilbert and Nadeau 2002; McKinnon 2006, 2007).

As shown in Table 2.1, within the time span assessed in this study, modal shifts occur from rail and water modes to the road mode in years 1973–2014; the USA displays the highest rail share in the listed countries; this provides the USA a strong competitive advantage. Japan and China show the highest fraction of IWW freight. These advantages of these modes are underexploited in terms of their complementarity and energy efficiency savings. For example, a logistics firm would need to consume 722 kcal per t-km on a truck in comparison with 58 for rail and 199 for IWW (Japan data: IIEJ 2010). This calls for greater expansion of energy efficient transport modes in future.

Table 2.1 Freight inland transport: modal split in selected years (excludes air cargo and pipelines)

Country	Mode	2010 (% share)	2014 (% share)
European Union (28 countries)[1]	Road	75,7	74,9
	Rail	17,4	18,4
	Inland waterways	6,9	6,7
	Total	100,0	100,0
China[2]	Road	31,1	32,1
	Rail	19,8	15,5
	Inland waterways	49,1	52,4
	Total	100,0	100,0
The United States of America[3]	Road	53,4	45,9
	Rail	35,9	42,5
	Inland waterways	10,7	11,6
	Total	100,0	100,0
Denmark[4]	Road	88,5	88,8
	Rail	n.a.	n.a.
	Inland waterways	11,5	11,2
	Total	100,0	100,0
Germany[5]	Road	70,6	71,3
	Rail	18,6	18,8
	Inland waterways	10,8	9,9
	Total	100,0	100,0
Japan[6]	Road	54,8	50,7
	Rail	4,6	5,1
	Inland waterways	40,6	44,2
	Total	100,0	100,0
The United Kingdom[7]	Road	89,0	87,0
	Rail	10,9	12,9
	Inland waterways	0,1	0,1
	Total	100,0	100,0

Sources: The EU data 27 (member states), for 2010, and Eurostat (2012, 2018), National Bureau of Statistics China (2018), U.S. Data in: Bureau of Transportation Statistics (2018). Statistics Japan (2018) and IIEJ (2010). UK data in: Eurostat (2018); Germany: Eurostat (2018); Denmark: Eurostat (2018)

The modal shift changes in road freight transport are correlated to the RE: The road mode will gain market share at the expense of the others modes; this higher mode increases (1) energy use, (2) oil import expenditures in Japan and (3) CO_2 emissions. The RE effect also impacts indirectly on the mix of modes that serve the freight market.

2.2.3 How Far Will Freight Travel and Its Effect on Energy Use

Tables 2.2 and 2.3 show that the average haul has varied from highest in the USA, Mexico and China to lowest in Japan and Denmark. Meanwhile, European nations (Germany included) are recording an increase in the average length of haul (Table 2.2). The UK registers a rise in the average distance that trucks travel. The only countries that register a fall in the average length of haul are Japan and the USA out of all the listed countries. Four developments explain this shift in Japan and the USA: first is the decline in the manufacturing base, second is the fall in the number of new warehouses, third is the higher dispersion of the warehousing sites across the territory and fourth is an increase in urban congestion both within Japan's and US cities which increases the frequency of trips.

Sustained increases in road freight demand have created a number of externalities (these are defined as 'the action of one person directly affects the prospects of another' (Stern 2009)—truck emissions in an urban setting) related to oil demand, local pollution, accidents and urban congestion (FreightVision project 2010).

One of the key externalities is freight energy use which increases oil dependence. In the USA, freight energy consumption (road mode only) is responsible for 23% of total energy use (petroleum only) of the entire transport sector (ORNL 2015). In contrast to that nation, China's freight energy use takes 30% of all transport energy use (Rail and road modes, National Bureau Statistics China (2018), whilst road freight transport takes 8% of total transport energy use. EU freight energy use accounts for 30% of total transport energy (Faberi et al. 2015) and for 35% in Japan's case in 2008 (IIEJ 2010). Germany's freight transport takes 26% (road freight sector) of the total transport energy (AGEB 2017). On average freight energy takes about a quarter of total transport energy use in many countries.

Road freight transport accounts for 20% of total transport energy use of UK transport (DECC 2016) and for 40% in 2010 in Denmark. These figures compare to 45% of freight transport (all modes) in total energy consumption of the world trans-

Table 2.2 Average haul for road freight transport

	1990 km	2010 km	Status of manufacturing sector
The USA	479	432	Declining manufacturing sector. Warehousing space grows (annual average 3.2% between 2002 and 2012).
Japan	89	54	Stable industrial activity until 1990s. Warehousing space is growing slowly year on year.
The UK	77	93	Declining manufacturing sector; growing geographical dispersion of warehousing. Warehousing grows slowly.
Denmark	48	64	Stable manufacturing sector. Warehousing space declines at 4.8% on compounded annual average growth rate in 2000–2014.

Source: See Table 2.1 for sources. U.K. data in DfT (2016), Statistics Denmark (various years). For Japan, Data Monitor (2011)

Table 2.3 Average length of haul in selected economies

Country	1990 (kilometres)	2010	2012	2015
European Union (28 countries)			120	124
The United States	479	432	421	n.a.
China	n.a.	n.a.	187	184
Germany	n.a.	n.a.	106	104
Japan	89	54	n.a.	48
The United Kingdom	77	93	98	95
Denmark	48	64	95	86
Mexico	237	518	n.a.	n.a.

Source: for EU countries: Eurostat (2017); Author's own calculations based on: Bureau of Transportation Statistics, U.S.A. (2018); National Bureau Statistics China (2018). Author's own calculations based on: Statistics Japan (2018); U.K. (DfT 2016); Denmark: Statistics Denmark (2016). Mexico: INEGI (2011). For Mexico 1990 column refers to the year 2003. The average length of haul is calculated by dividing tonne-km by tonnage, both in annual figures

port system (Sims et al. 2014, IPCC, 5th AR, 2). Freight energy is mostly consumed by trucks on a global scale and its scale is usually overlooked by many experts.

There are no official statistics on the share of energy consumption of the freight transport sector for Germany, Denmark or for Mexico. Transport as a whole is responsible for the largest proportion of oil consumption in Denmark in 2005 (Statistics Denmark 2007). In Denmark, road freight is responsible for 40% of transport CO_2 emissions, NOx emissions and those of particular matter (PM10).

In all sampled countries, it is the only sector that continues to increase its dependence on oil unlike the industry or household sectors which have decreased their dependence on that energy source. As in Denmark, China's and Japan's freight transport sectors take increasingly a larger share of total oil consumption.

Previous studies on rebound effects for road freight mode focus only on useful work (t-km) but not on distance travelled by trucks; we also assess the distance and the relationship among (1) distance travelled to move cargo (represented by truck km) and (2) fuel economy trends and (3) fuel prices. Distance has grown in all countries (Table 2.4), however, US- and China-based trucks travel the farthest, reflecting their territory and size of industrial prowess. Japan's trucks are travelling long distances and have similar per capita freight activity to Denmark's freight industry. Trucks based in the European Union (28 MS), travel shorter distances as do US-based trucks. German trucks travel far shorter distances than Japanese ones (business, private and mini sized vehicles) even though these two economies are global traders (Table 2.4). Germany-based trucks travel an average distance just above the level of UK trucks, whilst Mexico-based trucks travel shorter distances comparatively, although data for Mexico excludes smaller trucks and it is incomplete. The larger levels of distance travelled indicate falling transport costs in general.

Table 2.4 Country characteristics: vehicle kilometres travelled

	1990 Truck km (Million)	2010 Truck km (Million)	2016 Truck km (Million)	Comments
The USA[a]	235,303.3	461,021,9	442,503.9[a]	Fuel economy grew slowly (Heavy single unit and combined trucks).
European Union (28 MS)		107,813	116,858	Fuel economy (Excludes vans, pickups and road tractors). See term report of the EEA.
China[b]			422,630[b]	
Japan (Total)[c] Subtotal below:	226,064	204,923	201,294	Fuel economy improves (less Litres per tonne moved) but worsens in recent years.
Business vehicle	44,258	62,992	59,870	The nation is the top performer in our sample.
Private vehicle	96,358	66,667	65,548	The figures are the subtotal for Japan.
Mini-sized vehicle	85,420	75,264	75,876	
Germany[c]		29,300	30,083	Fuel economy improves significantly
The UK[c]	24,939.5	26,387	27,710 (2017)	Fuel economy gains are achieved in period.
Denmark	6884	9906	8234	Fuel economy gains occur slowly.
Mexico			3688	Fuel economy worsens. Data incomplete.

Source: Oak Ridge National Laboratory
[a]ORNL (2015) Bureau of Transportation Statistics (2018), U.S. Data: 2018;
[b]Year of observation: 2006, for China. Japan data 2017 includes: vans, pickups, lorries and road tractors (IIEJ 2017). Statistics Japan (2018). Germany: Eurostat (2016)
[c]U.K. Dft (2016, various years); Statistics Denmark (2018); includes data of vkm vans and trucks

The RE should increase distance (truck kilometres travelled) but the growth of the latter would be offset by the additional cost of fuel economy improvements of truck engines (Greene 2012). The evidence that we review shows that truck kilometres continue (distance travelled) to increase notwithstanding the cost of fuel economy technology.

US trucks travel longer distances than any other country except China. US trucks also travel much further than EU or Japan-based trucks. US-based trucks travel almost ten times as much as German ones; Mexico-based trucks travel far less than US ones perhaps because economic activity is highly clustered in the centre of Mexico. To conclude if the other economies catch up with the US level of freight movement this could become a powerful disincentive for achieving sustainable transport system. We now examine freight energy of economies for which data is available that enables analysis of rebound effects.

2.3   Material

In this section, we review studies of road freight demand from international studies.

2.3.1   Literature Review

Studies on the rebound effect with an application to freight transport are few. Recent studies include those for the USA (Winebrake et al. 2012; EPA 2011), for Portugal (Matos and Silva 2011) and for the EU by EC (2012).

Those works, however, exclude the case of Japan and of Denmark. A thorough review is presented in Sorrell and Dimitropoulos (2008) but it focuses mainly on other sectors (households, industry and passenger cars). Anson and Turner (2009) estimate rebounds for the Scottish freight transport sector (road); they find a 39% rebound effect of transport oil use using a general equilibrium model of the Scottish economy. These findings, however, cover the entire economy and cannot be empirically tested. Matos and Silva (2011), using a sophisticated econometric model, find a rebound effect of 24% for Portugal's road freight sector; work by Gately (1990) finds a rebound effect of 15–45% (in the long run) for road freight sector of the USA. However, the latter definition differs from that of Sorrell and Dimitropoulos (2008). Winebrake et al. (2012) review studies on the US freight transport sector, and the growing global literature, on the rebound effect for this sector. The cited authors acknowledge two problems with the literature: (1) the lack of a systematic analysis and of (2) the lack of harmonization of metrics that do not allow a common understanding of the RE.

A White Paper from the American Council for an Energy Efficient Economy (Nadel 2012) reports around 100 studies on the rebound effect. It finds that most studies are based on the (private) vehicle travel studies. A study by the EPA (2011) finds a rebound effect of 10% for private vehicles.

The following studies can be used as approximations of estimates of the indirect RE. These approximations have been discussed in Agnolucci and Bonilla (2009). The cited authors say 'Many freight studies arrive at widely different conclusions reflecting the heterogeneity of economies, the spatial concentration of production and consumption points, and warehousing practices. Some national economies are made up of larger agricultural and manufacturing or service sectors. As a result studies of freight demand (or energy consumption of freight demand) tend to produce widely ranging estimates of price and income elasticities'. Thus, when comparing estimates on RE, analysts should be aware of the fundamental differences between models and datasets.

The following elasticities can be interpreted as implying that a price decline of x percent level is associated to an increase of x percent useful work (t-km of road freight demands). A diverse set of cost elasticities, for the US economy, was found in Winston (1981), i.e. from −0.14 to −2.96 (the mode is −0.29; mean is unity) across the set of 12 commodities where road freight demand had been estimated.

At the aggregate level, a widely cited study (Oum 1989) focused on the North American freight sector and reported a variety of price elasticities for road freight. Oum (1989) tests various functional forms to estimate elasticities of road freight transport. Using a price index reflecting the length of haul, Oum (1979) found that price elasticities range from −0.045 to −0.15 over several years.

An extensive review on freight transport studies for Great Britain (DfT 2001) reported a price elasticity of 10% – for freight demand (length of haul). Agnolucci and Bonilla (2009) use data on gross value added of industry and find a price elasticity of 20% (British road freight only) and an income elasticity value of 65% for economic activity (income elasticity of road freight). Using, however, a GDP and imports as explanatory variables, produces an elasticity value of 90%.

In a highly regarded review of price elasticities of the road transport, Graham and Glaister (2004) report a mean of the elasticities of −1.07 which is obtained from values ranging from −7.92 and 1.72. There is no exact agreement on the size of price elasticities according to the cited authors.

The above studies can only be taken as approximations of the direct rebound effect according to the literature. In theory, this wide variation in price elasticities should produce different values of the rebound effect depending on the commodity type (Matos and Silva 2011).

2.3.2   Data Inputs for the Rebound Effect

Table 2.5 describes the data input to build the estimates for the rebound effect of the road freight transport sector. The data gives the median, standard deviation and the maximum values. The most volatile variable is GDP.

2.3.3   Theoretical Models on the Rebound Effect

Following the mathematically derived analysis in Sorrell and Dimitropoulos (2008), Matos and Silva (2011), (Khazzoom 1980) and Greene et al. (1999), we work out the following equations to estimate the rebound effect in a dynamic setting using time series data sets for the three countries.

We estimate the RE by using definition 2 based on national published statistics. Following Sorrel (2009) the two relevant definitions that we use are:

1. Definition 1 represents the elasticity of demand, in a year, of useful work with respect to the price of useful work.
2. Definition 2 represents the demand of useful work, in a year, with respect to annual changes in energy efficiency of the freight transport sector.

Table 2.5 Summary of data for the econometric analysis

Variable name and units	Median	Standard deviation	Maximum
Diesel prices (Denmark) Danish Krona per Litre	693.37	420.99	1612
Oil prices (Japan) (US$/bbl)	22.6	19.53	100.9
Gasoline price Japan (Yen/Litre)	124		
Diesel prices (US$/Litre) (using purchasing power parity)	0.94	0.41	1.9
Freight transport performance (Units: billion t-km) Denmark	9.19	2.16	11.80
Freight transport performance (billion t-km) Japan	203.88	83.46	355.04
Freight transport performance (billion t-km) UK	216	41.76	276.8
Truck-km Japan (business owned, private vehicle and mini vans) (billion vkm)	196.609	68.28	267.13
Truck-km Denmark (million vkm; heavy goods vehicles)	1452.2	176.7	1743
Truck-km U.K. (billion vkm; heavy goods vehicles)	24.57	3.6	29.3
GDP (Japan) Units: Constant US$	4095.8	117758.7	
GDP (Denmark) Units:	139.07		
GDP (UK) billion Constant prices	1202.6	3.53+e11	1.81+E12
Fuel consumption (diesel fuel, road freight transport) Japan (000's TOE)	25878	3410.13	28153
Fuel consumption (road freight transport) U.K. (million tonnes of fuel)	7	1.2	8.09
Fuel consumption (road freight transport, PJ) Denmark	19.45	4.53	27.16

Source: Gasoline Prices in IIEJ (various years); freight transport activity in IIEJ; Statbank Denmark (2013) for 1980:2011. UK DECC (2016–various years). See Fig. 2.1 for further sources of data

2.3.4   Regression Analysis of the Rebound Effect

In Eq. 2.1, we test for the rebound effect for the cited countries, through the following econometric model based on time series data for years: 1978–2008. This is Definition 2 in Sorrell (2008) as we discussed above.

Consider the following notation:

tkm = tonne km per capita per year
p = cost of useful work
LF = tonne carried per kilometre travelled
L/km = Litres per kilometre travelled
ln GDP = logarithm of GDP
ln oilp = logarithm of oil prices

Subscript t: time, 1980–2008,

$$\ln \text{TKM}_t = \ln \beta_0 + \beta_1 \ln \dfrac{P_t}{LF_t \Big/ \dfrac{L}{km_t(\text{trucks})}} + \beta_2 \ln(\text{GDP}_t) + \beta_3 \ln(\text{oilp}_t) + \text{Error}_t \qquad (2.1)$$

Equation 1 allows the estimation of definition 2 (Sorrel 2009) of the rebound effect. In Sect. 2.4, the results are discussed.

2.4   Discussion on Rebound Effects and Freight Transport

Using the estimated parameters of the econometric model (Eq. 2.1), we estimate the direct rebound effect of road freight following the definitions of Sorrell (2008) and Matos and Silva (2011).

2.4.1   Discussions

In this section, we discuss how we obtain the results and why the results are important. We discuss (1) the RE and (2) interpret the econometric results of using equation for the three countries sampled. The econometric estimation is implemented using the logarithms of the data described in this section. All equations were estimated using SPSS for years 1980–2008.

Table 2.6 tabulates the values for two sample periods (1980–2008; 1986–2005) of the rebound effect for the three countries. In Table 2.6, we omit the rest of the estimated values for the sake of summarizing the analysis. The rebound effects are calculated using Eq. 2.1 and these values are given after estimating the econometric equation based on time series data specific to each of the three countries. Table 2.7 lists the actual elasticities for distinct periods (1980–2008), 1986–2005.

We use definition 2, as defined above, to estimate the rebound effect of the road freight sectors. These effects for different periods are estimated for both Japan, Denmark and the UK.

The econometric results (Table 2.6) reveal the RE rises over time: the estimates of 1980–2008 are lower in absolute terms than those of 1986–2005. In other cases the RE estimates behave otherwise: Japan's RE estimates or Denmark's show this.

The size of the RE can be sensitive to the sample period, and so the model is tested over a two different periods (1980–2008 and 1986–2005); this procedure should reveal or confirm, or even reject, the appearance of the RE over time for the three countries. The period (1980–2005) should reveal the effect of high oil prices in the early 1980s and in the 2000s; and the 1980–2005 reflects the effect of lower oil prices on fuel economy and on energy saving practices of truck freight transport.

Table 2.6 Econometric results for the direct rebound effect in 1986–2005

(Dependent variable: (t-km)	1980–2008 Estimated coefficient (β₁, Eq. 4) Direct rebound effects (A)	1970–2008 (β₁, Eq. 4)	1986–2005 (β₁, Eq. 4) Direct rebound effects (B)
Japan	0.14** R squared: 0.97		−0.014 R squared: 0.99
			(−0.01;with lag dependent variable) R squared: 0.98
The United Kingdom	−0.25** R squared: 0.97	−0.10 R squared: 0.95	−0.67* R squared: 0.90
	−0.089** (with a lag dependent variable)		
Denmark	−0.41** R squared: 0.97		−0.11** R squared: 0.99
			−0.013 (with a lag)
	R square 0.97		

Notes: ** statistically significant at 5% level. See equation dependent variable: t-km per year (road freight only). *RE* rebound effects

Table 2.7 Elasticity for freight energy use: subsample, non-direct rebound estimate

Rebound effects	1980–2008 Elasticity of diesel use after % change in energy efficiency (based on Eq. 2) (C)	1986–2005 Elasticity of diesel use after % change in energy efficiency (based on Eq. 2) (D)
Japan	−1.14	−0.98
The United Kingdom	−0.75	−0.23
		−0.91
Denmark	−0.59	−0.89

To calculate the RE using Eq. 2.1, the following decision rule is used. If the RE is larger than 1, the effect of changes in energy efficiency on energy use (road freight) will be more than proportional and negative. If the value of the RE is less than 1, the effect of energy efficiency on energy use will be positive: that is energy use of freight transport will increase rather than decrease as the price of energy services declines. In the absence of RE, the value should be 1: increase in energy efficiency should match the decrease in freight energy. The RE estimates are given in Table 2.6.

The direct RE estimates (Table 2.6 columns A and B) are based on econometric estimates for the three countries. With the exception of the UK, these RE estimates are relatively small if the estimation is based on the 1980–2008 (Table 2.6, Column A) sample.

In this period, the increase in the energy cost is associated to an increase in useful work (t-km) of Japan's freight transport (positive rebound effect). An increase in energy efficiency of truck freight by 1% induces a reduction of 1.14% in energy use (if the sign was negative). Note the RE has a value with a positive 14%. Similarly, for the 1980–2008 if the elasticity of useful work with respect to (shortened to w.r.t.) fuel cost per tonne is 0.14 (Table 2.6, column A), the elasticity of diesel consumption w.r.t. fuel efficiency can be estimated as: −1.14.

The analysis for Japan's RE mirrors that of Denmark (Table 2.6, Column A). Denmark's freight transport sector is sensitive to the behaviour of energy efficiency of the sector too. The value (Table 2.6, column 'A') of the direct rebound effect is (−0.41) or 41%. If the elasticity of useful work w.r.t. fuel cost per tonne is −0.41 (Table 2.6, column A), the elasticity of diesel consumption w.r.t. fuel efficiency can be estimated as: −1.41. The Danish case shows negative RE effects: higher energy costs are associated to lower useful work.

Using the sample 1986–2005 (Table 2.6, Column B) shows a different picture: none of the RE estimates lies above unity; this reflects that energy cost reduces useful work, whilst two of the RE estimates are negligible: −1.3% and −1.3% (Japan and Denmark with a lag coefficient). It appears that periods of low oil prices are associated to higher RE effects in absolute values.

One can further interpret the results of Table 2.6 as follows: e.g. For the UK case (Table 2.6, column 'A'), if the elasticity of useful work w.r.t. the fuel cost per tonne is −0.25, the elasticity of diesel demand w.r.t. energy efficiency can be estimated as −0.75, using our definition 2 of useful work. Thus, the size of the direct rebound is 25%. This means that if fuel efficiency improves by 10% diesel demand will fall by 7.5 % (not 10% and hence this is not a proportional change) in the period of estimation (1980–2008). Without the RE effect diesel demand should decline by 10% following energy efficiency gains.

After comparing the estimated equations for the three countries (Table 2.7, Columns 'C' and 'D'), the analysis shows that the magnitude of the calculated elasticity values differs strongly among countries. The incentives for energy saving practices falls in periods of low oil prices and rises in periods of high oil prices. The latter period is associated to a more rapid adoption of the fuel efficient engine in Japan and perhaps less so in the UK. The adoption of efficient trucks allows Japan to expand output (t-km) at the expense of energy costs which explains the positive coefficient of Japan's estimates out of all three countries (column 'B'). Acquiring more fuel efficient trucks translates as more capital for energy.

We can compare our estimates of Tables 2.6 and 2.7 to others. For the UK Barker and Foxon (4CMR Center 2006), a 11% macroeconomic rebound effect has been reported; for transport they estimate 4–7% levels but they use a different definition of the effect and employ mileage driven to measure the effect amongst other macroeconomic factors. Most likely, the RE ranges between 10 and 30% (Sorrell 2007).

The size of the RE is likely to be proportional to the share of energy costs of the total energy costs of the energy service (the movement of freight). The price of diesel (US$/Litre) is more expensive in the UK than in Japan or Denmark (IEA 2012, Energy Prices and Taxes) which increases the cost of the energy service in the UK significantly. The high cost of that energy service should be associated to a rapid adoption of truck engines and a strong RE but the estimates on RE (Table 2.6) show otherwise. In absolute terms, Japan's RE estimates are the largest out of all three countries.

Diesel costs are important for freight transport demand, and for plant location decisions. As far as Denmark, the downward trend in diesel costs also explains the growth in truck freight activity (Fig. 2.1). This is the main price that road freight firms need to pay.

2.5   Rebound Effect and Fuel Economy

This section discusses whether fuel economy gains, of goods vehicles, influence the size of the RE. The size of the RE will also vary possibly by sector.

2.5.1   Fuel Economy and Fuel Use of Freight Transport: 1980–2006

Rebound effects can also emerge as a result of lower costs of new vehicles (or fuel consuming equipment), that is lower capital costs. This translates into lower vehicle running cost as the fuel efficient truck fleets are renewed. This process gives expenditure savings to firms; and this extra expenditure is channeled to purchases of new vehicles, which in turn, can slow down the improvement in fuel economy on the road.

Figure 2.4 shows actual (on-road) fuel economy of trucks (L/100 km) 1980–2010 for our three economies. This period is dominated by stable diesel (and gasoline) prices (Figs. 2.5 and 2.6). Fuel economy gains (fewer L/100 km) do respond to rising diesel prices which doubled in 2000 onwards in the entire OECD region (Figs. 2.5, 2.6). Japan's diesel prices also doubled in the same period. The top performing countries as far as fuel economy is concerned are Japan and Denmark and show the most efficient fleets in terms of overall energy efficiency and fuel economy. These fuel economy levels of the two countries are likely to have continued in the period not shown in Fig. 2.4, i.e. 2010–2015. Figure 2.7 does show a decline in fuel economy for Danish trucks larger than 6 tonnes but the data excludes vans. In general fuel economy levels do show a downward trend regardless of changes in diesel prices which is explained by technology change, i.e. better engines.

The gains can be seen (less L/100 km travelled) for the UK and Denmark in Figs. 2.4 and 2.7. The slope of fuel use per kilometre for the Japanese and the Danish cases does not decline as steeply as the UK does in 1980–2010. Japan shows a worsening of the ratio of fuel economy (more Litres per 100 km) but it is already far more fuel efficient that other countries. Danish fuel economy, in 1990–2005, improves by almost 34% (Fig. 2.4) for 6 tonne gross vehicle weight. The rebound

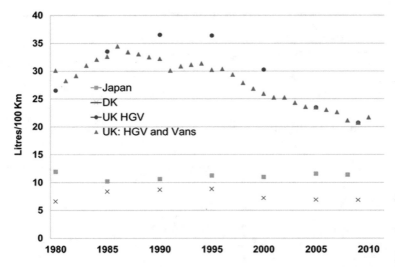

Fig. 2.4 Actual (on-road) fuel economy of trucks (L/100 km). Data for UK includes HGV and vans (<2.6 tonnes vehicles); for Denmark, HGVs and vans; and for Japan HGV's, vans and mini-vans. Elaborated by the authors based on Danish Energy authority (2016), IEJ (2011); DfT (2011), and UK: DECC (2013)

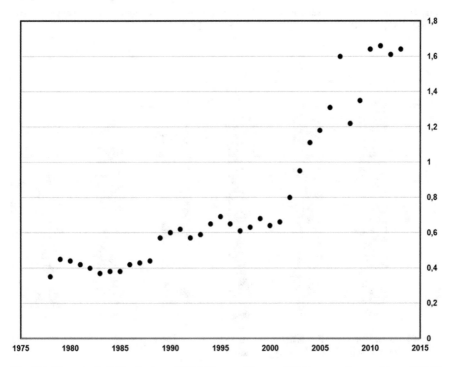

Fig. 2.5 Time trend of diesel prices: OECD Europe. (Source: IEA Energy prices and taxes (US $/ unit using Purchasing Power Parity)

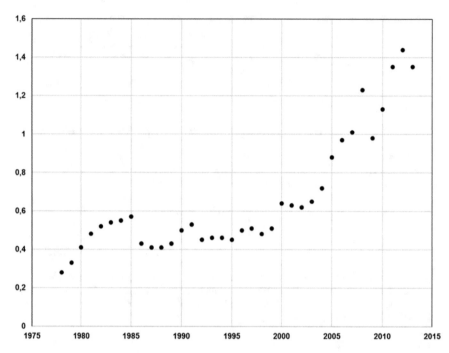

Fig. 2.6 Time trend of diesel prices: OECD Total. (Source: IEA Energy prices and taxes (various years). Total price (US$/unit using Purchasing Power Parity))

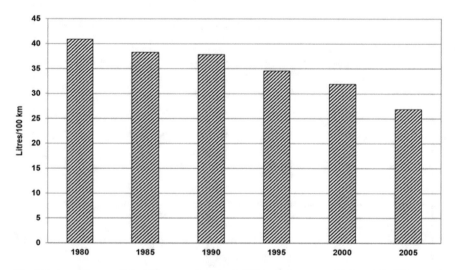

Fig. 2.7 Actual (on-road) fuel economy of trucks (L/100 km) >6 tonnes vehicle weight. Elaborated by the authors based on Danish Energy authority (2016)

effect explains why levels of on-road fuel economy have not declined as much as it is desired for the UK case. Figure 2.5 depicts the timeline of diesel prices within OECD Europe and Fig. 2.6 OECD total; it reveals an upward trend since 2000: prices doubled in 2005–2015, but in general, the distance travelled by trucks is still increasing.

Fleet wide fuel economy of truck freight (L/100 km) depends on: the freight vehicle mix (large or small trucks and delivery vans); (b) the fuel mix of the freight vehicle stock (diesel, gasoline, biofuels and diesel hybrid) and (c) the vehicle age of trucks. Other factors include:

- Distance travelled (short or long hauling distance) which is associated with average speeds. Or optimal speeds of truck driving;
- Congestion level: this affects the level of average speed and increases the frequency of cold starts of truck engines;
- Other technical measures (vehicle design and aerodynamics).

2.5.2 Fuel Intensity of Moving Freight by Sector

Despite improvements in overall average truck fuel economy (large truck only, excluding the van sector) and despite the fall the intensity of total Danish freight diesel use increased by 215% between 1980 and 2006 (truck only). In Japan, diesel use in the sector rose by 150% (1965–2008) and in the UK by 84% (1980–2011).

Besides truck fuel economy, discussed above, overall energy intensity of truck freight (MJ/t-km) is influenced by:

- Vehicle utilization (tonnes per truck, or loading level of individual vehicles);
- Average hauling distance (t-km divided by tonnes lifted);
- Type of freight moved (bulk goods, manufactured);

2.6 Conclusions

This is the only analysis that uses evidence from country regions from separate continents to study the trends and changes of the freight transport sector during 1950–2016. In this chapter, we examine trends in t-km for the USA and other leading economies and find that the USA shows the smallest increase in road freight in the examined economies. Japan registers the largest decline in road freight transport and China shows that largest increase of activity of that sector. The rate of growth of 1960s Japan is being mirrored by China in recent years. The modal split shows that China has a more balanced modal distribution for moving freight than the competing nations and the balanced modal distribution is likely to have avoided an even larger increase of road freight flows. The EU and Germany register growth just above the US and Mexico levels while the UK registers one of the smallest rates of increase in our sample.

We identify five key factors that determine freight transport activity: population density, trade, how far trucks travel, their average length of journey and the degree of competition among the transport modes.

The examination of average length of haul finds that China and Mexico are both catching up with levels of distance travelled by US-based trucks. The USA shows the highest level of this metric followed by China. The length of haul in most economies is growing and not decreasing which signals a need for a more sustainable freight transport system.

Trucks are travelling longer distances more than ever in most of the economies examined in the period considered. Longer distances imply that transport costs are relatively low in most countries; however, countries such as China will continue to subsidize diesel prices, while Europe will continue to tax diesel fuels much more than the USA and Mexico currently do. The end result of the subsidy is that the distance travelled by trucks will continue to grow in China and less so in Europe and in the USA.

Worryingly we find that the distance travelled by the US- and Mexico-based trucks are getting longer which produces even more congestion on highways and cities. China's freight transport sector is also registering long distances for trucks which is likely to stimulate energy use of that sector.

The cross country comparison shows that the per capita level of freight moved domestically by road in many countries is quickly catching up with the US level. The latter is the top performer in terms of freight moved per capita. The freight intensity of economies is highest in the USA historically; however, convergence among key industrial countries is occurring fast.

It is also the first analysis of three countries to estimate the rebound effect using historically observed data which increases the reliability of the results. A number of conclusions can be drawn from assessing the empirical evidence (a) on the rebound effect for the four countries.

Having sketched out the differences across key economies as far as the sector, we compare how industry tends to use trucks depending on how extensive the territory is. On the one hand, the vast territories of China, the USA and Mexico impact on the length of haul, truck size and the volume of cargo; and on the other hand, the limited land available in Japan, the UK and Denmark as well as the EU imply these nations must have similar levels of the length of haul, of distance travelled levels, of truck size. The analysis reveals inclusion of geography variables help explain freight transport flows.

The direct rebound effect, estimated through time series data, allows a cross country comparison of the energy efficiency of moving cargo by truck, and of the effect of fuel costs on freight movements.

The analysis presented here relies on indirect measures of the rebound effect. The analysis, however, does capture a general pattern: the RE rises and falls but only slightly over time. This confirms the findings of small and Van Dender (2004) who also found a declining rebound effect for the case of US private cars. The freight transport estimates on rebound levels are in the range of 7–25%. These findings, at least for the UK, also echo those of Matos and Silva (2011) who reported 24% direct rebound estimate for Portugal.

The analysis for the three countries shows that since 1978–2008 the rebound effect has been greatest in the UK. The gains in fuel economy of 1% leads to a decrease in fuel consumption of diesel by 0.7% for the UK. This is not a proportional decrease in fuel consumption because the rebound effect is 29% for this nation. The Japan's and Danish estimates, on the rebound effect, are high considering the volumes of diesel fuel that the freight sectors consume annually.

The incentives for energy saving practices falls in periods of low oil prices and rises in periods of high oil prices. The latter period is associated to a more rapid adoption of the fuel efficient engine in Japan. The adoption of efficient trucks allows Japan to expand output (t-km) at the expense of energy costs which explains the positive coefficient of Japan's estimates out of all three countries. Acquiring more fuel efficient trucks translates as substituting more capital for energy.

In addition to the analysis of the RE, we have observed that both on-road fuel economy (large truck-vans) and the carbon intensity of truck freight (large trucks and vans) continue to fall in the USA and the UK, but not in the case of Japan or China.

The total energy use of trucks will continue to rise because of the absence or lax fuel economy regulations, a low average vehicle load and a larger hauling distance in the leading economies. The popularity of vans will continue worsening fuel economy performance of the entire road freight sector for all economies studied.

Future work should assess the RE of freight transport at the level of commodity but this requires detailed data on freight transport flows which is not normally publicly available. Notwithstanding these limitations, we have produced estimates on how entire freight sectors are likely to choose to buy freight transport services.

Policy measures should encourage a shift to alternative transport modes (which requires an integrated transport policy), the adoption of alternative vehicle technologies and the improved utilization of trucks, as well as the adoption of best practice programmes for vans.

Acknowledgments Research assistance for updating Graphs and Tables is gratefully acknowledged by J. Navarro Guevara. Research assistance by Nihan Akyelken (University of Oxford, Transport Studies Unit, School of Geography and Environment) is also acknowledged.

Bibliography

ACEEE (2012) The rebound effect: large or small? Report prepared by Steven nadel. An ACEEE white paper, pp 1–9

AGEB - AG Energiebilanzen e.V. (2017) Evaluation tables for the energy balance for Germany: 1990-2016, Berlin. https://agenergiebilanzen.de/index.php?article_id=29&fileName=ausw_07082017ov_engl.pdf. Accessed Oct 2017

Agnolucci P, Bonilla D (2009) UK freight demand: its elasticities and decoupling. JTEP 43(3):317–344

Anson S, Turner K (2009) Rebound and disinvestment effects in oil consumption and supply resulting from an increase in energy efficiency in the Scottish commercial transport sector. Department of Economics, University of Stratchclide, Glasgow. Working paper no. 09–01

Barker T, Kohler J (2000) Charging for road freight in the EU: economic implications of a weigh-in-motion tax. J Trans Econ Policy 34(3):311–331

Bjorner T (1999) Environmental benefits from better freight transport management: freight traffic in a VAR model. Transp Res Part D Transp Environ 4(1):45–64

Brookes L (1990) The greenhouse gas solution fallacies in the energy efficiency solution. Energy Policy 18(2):199–201

Bureau of Transportation Statistics (U.S.) (2018) Various years. Data: freight inland transport: modal split in selected years. https://www.bts.gov/. Accessed June 2018

Cambridge Center for Climate Change Mitigation Research (4CMR) (2006) The macroeconomic rebound effect and the UK economy. Report prepared by the 4CMR, Department of Land Economy, University of Cambridge, with Cambridge Econometrics Ltd; Policy Studies Institute (PSI) University of Westminster, Dr Horace Herring, Open University. Final report to DEFRA, pp 1–98

China Federation of Logistics and Purchasing (2018) Data: vehicle kilometers of trucks. http://www.chinawuliu.com.cn/english/. Accessed Feb 2018

Danish Energy Authority (2016) Energy statistics, Denmark. https://ens.dk/en/our-services/statistics-data-key-figures-and-energy-maps/annual-and-monthly-statistics. Accessed Mar 2016

Data Monitor (2011) Japan Logistics and Express Outlook. An analysis of key trends driving the Japanese logistics sector. Data: no. of warehouses. www.datamonitor.com. Accessed Apr 2015

DECC (Department of Energy and Climate Change) (2016) Data on vehicle t-kms vans, gasoline consumption. https://www.gov.uk/government/collections/digest-of-uk-energy-statistics-dukes. Accessed Mar 2016

Deustche Bank (2015) Logistics in Germany: only modest growth in the near term. Current issues sector research report by E Heymann, pp 1–12. https://www.dbresearch.com

DfT - Department for Transport (2001) Integrated transport and economic appraisal. Review of freight modelling, report B1. Review of GB freight models, June 2002, pp 1–110

DfT - Department for Transport (2016) Transport statistics Great Britain, London. https://www.gov.uk/government/collections/transport-statistics-great-britain. Data on t-km. Accessed Jan 2017

EPA - United States Environmental Protection Agency (2011) Final rulemaking to establish greenhouse gas emissions standards and fuel efficiency standards for medium and heavy duty engines and vehicles: regulatory impact analysis. https://nepis.epa.gov/Exe/ZyPURL.cgi?Dockey=P100EG9C.TXT. Accessed Mar 2011

European Commission (2006) European Energy and Transport: trends to 2030, update 2005. https://europa.eu.int/comm./energy_transport/en/lb_en.html

European Commission (2012) Addressing the rebound effect. A project under the framework programme. Env. G.4/FRA/2008/0112

Eurostat (2012) EU transport in figures: statistical pocketbook 2012 report, pp 1–133. Brussels Belgium

Eurostat (2017) Energy, transport and environment indicators: 2017 edition. https://ec.europa.eu/eurostat/documents/3217494/8435375/KS-DK-17-001-EN-N.pdf/18d1ecfd-acd8-4390-ade6-e1f858d746da. Accessed Mar 2015

Eurostat (2018) EU transport in figures: statistical pocketbook 2018 report, pp 1–133. Brussels

Faberi S, Paolucci L, Lapillonne B, et al (2015) Trends and policies for energy savings and emissions in transport. Report, ODYSSEE-MURE project coordinated by ADEME, pp 1–83. Brussels, Belgium 2015

Freightvision (2010) Freight transport foresight to 2050 (2008-2010). http://www.freightvision.eu. Accessed Sept 2015

Gately D (1990) U.S demand for highway travel and motor fuel. Energy J 11(3):59–73

Gilbert R, Nadeau K (2002) Decoupling economic growth and transport demand: a requirement for sustainability. Paper presented at the conference on transportation and economic development, Transportation Research Board, Portland, Oregon, May 2002

Goldstein D, Martine S, Roy R (2011) Are there rebound effects for energy efficiency. An analysis of Empirical Da, internal consistency and solutions. http://electricitypolicy.com

Graham D, Glaister S (2004) Road traffic demand elasticity estimates: a review. Transp Rev 24(3):261–274

Greene D (2012) Rebound 2007: analysis of U.S. light-duty vehicle travel statistics. Energy Policy 41:14–28

Greene DL, Kahn JR, Gibson RC (1999) An econometric analysis of elasticity of vehicle travel with respect to fuel cost per mile using the RTEC survey data. Oak ridge National Laboratory, Oak Ridge

Hao H, Feiqi L, Zongwei L et al (2017) Measuring energy efficiency in China's transport sector. Energies 10(5):1–18

IEA - International Energy Agency (1997) The link between energy and human activity. IEA, Paris

IEA - International Energy Agency (2002) Transportation projections in OECD regions. IEA, Paris

IEA - International Energy Agency (IEA) (various years) Energy prices and taxes. Automotive diesel prices for commercial use in USD/Litre, Paris. https://www.iea.org/classicstats/related-databases/energypricesandtaxes/. Accessed Nov 2016

IIEJ (Institute of Energy Economics Japan) (2010) Handbook of energy economics and statistics, EDMC. The Energy Conservation Center, Tokyo

IIEJ (Institute of Energy Economics Japan) (2017) Handbook of energy economics and statistics, EDMC. The Energy Conservation Center, Tokyo

INEGI - Instituto Nacional de Estadística y Geografía (2011) Encuesta Anual de Transportes 2011, Aguascalientes. http://www.inegi.org.mx. Accessed Sept 2016

Jevons WS (1865) The coal question. Macmillan, London

Kamakate F, Schipper L (2009) Trends in truck freight energy use and carbon emissions in selected OECD countries: 1973-2005. Energy Policy 37:3743–3751

Khazzoom JD (1980) Economic implications of mandated efficiency standards for household appliances. Energy J 1(4):21–40

Koomey J (2011) A fascinating encounter with advocates of large rebound effects. http://www.koomey.com/post/3286897788

Loo BPY, Banister D (2016) Decoupling transport from economic growth: extending the debate to include environmental and social externalities. J Transp Geogr 57(C):134–144

Maier SC (2006) Among empires. American ascendancy and its predecessors. Harvard University Press, Massachusets

Matos JF, Silva FJF (2011) The rebound effect on road freight transport: empirical evidence from Portugal. Energy Policy 39(5):2833–2844

McKinnon A (2006) The decoupling of road freight transport and economic growth trends in the UK: an exploratory analysis. Paper presented at the Logistics Research Center, Heriot Watt University, Edinburgh, UK, Oct 2006

McKinnon A (2007) Decoupling of freight transport and economic growth trends in the UK: an exploratory analysis. Transp Rev 27(1):37–64

Nadel D (2012) The rebound effect: large or small. Report prepared by Steven nadel. An ACEEE white paper 1:9

National Bureau of Statistics China (2018) Statistics on Average Haul. Years:1990, 2010, 2012, 2015. http://www.stats.gov.cn/english/ . Accessed Jan 2018

OECD Data - Organization for Economic Cooperation and Development (2018) Freight transport, Paris. https://data.oecd.org. Accessed Sept 2018

ORNL - Oak Ridge National Laboratory (2015) Transportation Energy Data Book, Tennessee. https://cta.ornl.gov/data/index.shtml Accessed Dec 2016

Oum TH (1979) Derived demand for freight transportation and intermodal competition in Canada. J Trans Econ Policy 13:149–168

Oum TH (1989) Alternative demand models and their elasticity estimates. J Trans Econ Policy 23:163–187

Owen D (2010) The efficiency dilemma: if our machines use less energy, will we just use them more?. The New Yorker, Dec 20 & 27

Schipper L, Grubb M (2000) On the rebound? Feedback between energy intensities and energy uses in IEA countries. Energy Policy 28:367–388

Schipper L, Steiner R, Meyers S (1992) Trends in transportation energy use 1970-1988: an international perspective. Lawrence Berkeley Laboratory, pp 1–31. University of California. https://escholarship.org/uc/item/9871n2zk. Accessed Sept 2014

Schipper L, Scholl L, Price L (1997) Energy use and carbon from freight in ten industrialized countries: an analysis ofr trends from 1973–1992. Transp Res Part D: Transp Environ 2(1):57–76

Sims R, Schaeffer R, Creutzig F et al (2014) Climate change 2014: mitigation of climate change. Transport (2014) (Chapter 8). https://www.ipcc.ch/site/assets/uploads/2018/02/ipcc_wg3_ar5_chapter8.pdf. Accessed June 2016

Sorrell, S (2007) The rebound effect: an assessment of the evidence for economy-wide energy savings from improved energy efficiency. UK Energy Research Center Report, pp 1–108. http://www.ukerc.ac.uk/publications/the-rebound-effect-an-assessment-of-the-evidence-for-economy-wide-energy-savings-from-improved-energy-efficiency.html. Accessed Jan 2015

Sorrell S, Dimitropoulos J (2008) The rebound effect: microeconomic definitions, limitations and extensions. Ecol Econ 65(3):636–649

Statistics Denmark (2007) Product statistics for shipping agents in 2006, Copenhagen. http://www.statbank.dk/statbank5a/default.asp?w=1280. Accessed Mar 2011

Statistics Denmark (various years) National transport of goods by type of vehicle, Copenhagen. http://www.statbank.dk/statbank5a. Accessed Mar 2011

Statistics Japan (2018) Japan statistical yearbook. Chapter 12. Transport and Tourism. Traffic volume by type of transport, Tokyo. http://www.stat.go.jp/english/data/nenkan/1431-12.htm. Accessed Jan 2018

Stern N (2009) A blue print for a safer planet: how to manage climate change and create a new era of progress and prosperity. The Bodley Head Press, London

TSGB (2013) Transport Statistics Great Britain. Transport statistics published In DfT (2016)

UK Institute of Mechanical Engineers (2013) Introduction to freight transport. http://www.imeche.org/knowledge/themes/transport/freight. Accessed Jan 2015

Van Dender K (2004) The effect of improved fuel economy. On vehicle miles traveled: estimates using U.S. State Panel Data. Paper presented at the Economics Colloquium, University of California, Irvine, pp 1–33

Winebrake JJ, Green EH, Comer B et al (2012) Estimating the direct rebound effect for on road freight transportation. Energy Policy 48:252–259

Winston C (1981) Disaggregate model for intercity freight transportation. Econometrica 49(4):981–1006

World Bank (2009) Reshaping economic geography. World Development Report 1-369, Washington DC

World Bank (2013) World Development Indicators, Washington DC. https://datacatalog.worldbank.org/dataset/world-development-indicators. Accessed Jan 2016

World Bank (2018) World development indicators, Washington DC. https:// datacatolog.worldbank.org/dataset/world-development-indicators. Accessed Jan 2018

Chapter 3
Europe's Sustainable Road Freight Transport to 2050: Closing the Gap Between Reality and Vision

3.1 Introduction

Freight transport activity of the European Union (EUFT) is a fast growing sector (road, rail and inland waterways modes) with a wide variety of actors, each of whom adhere to various paradigms and beliefs. EUFT activity is measured by commodity flows or vehicle movements (Nuzzolo et al. 2015). EUFT activity of the 28 member States has doubled in 1970–2012 and it is believed to double again by the year 2050 (Eurostat 2014; FVision 2010a, b). The sector has grown faster than the passenger sector (Eurostat 2014). A common freight strategy and the policies associated with it need to be established to control the growth of, and impacts of, freight transport. A backcasting exercise to the year 2050 can help achieve this. A series of studies on freight (DHL Deutsche Post 2012; U.S. Transportation Research Board 2013; Hao et al. 2015; WEC 2007) have produced explorative scenarios on the future of freight transport; but these studies are based on the assumption that the future of the freight transport sector is open: various freight pathways are possible.

Scenarios can be of two kinds, a scenario is defined as 'hypothetical sequences of events for the purpose of focusing attention on causal processes and decision points" (Kahn and Wiener 1967). In contrast to scenarios the practice of backcasting is defined as the "major characteristic of backcasting is a concern not with what futures are likely to be but with how desirable the energy futures can be attained' Robinson (1982). This chapter fills the gaps in the literature by imagining the future of sustainable freight by using a participatory approach for the sector through strategic conversations with stakeholders from entire supply chains operating in the European Union. A set of recommendations are formulated that rest on the limits to five dimensions: (1) economic growth; (2) resource use; (3) carbon emissions; (4) volume moved and (5) distance travelled by trucks as well as shifts in diesel use to other non-diesel energy.

© Springer Nature Switzerland AG 2020
D. Bonilla, *Air Power and Freight*, SpringerBriefs in Energy,
https://doi.org/10.1007/978-3-030-27783-3_3

This chapter has two goals: (1) to build a vision for the EUFT sector to 2050, by allowing stakeholders to imagine the future of freight (the movement of goods) by using the backcasting method (defined below) and (2) to produce a set of policy solutions to achieve the desired vision for EUFT to reach targets.[1] The targets focus on energy efficiency, carbon emissions, transport performance (t-km) and other fields. The policies proposed enable the EUFT sector and its stakeholders to reach the vision of 2050 through a set of policy actions that need to be implemented by 2050. These policies should take into consideration the argument of the limits to growth within a carbon constrained world (Jackson et al. 2014; Banister et al. 2011; Meadows 1972) and thus freight transport is no exception. The problems facing the freight sector have not changed since 2010: strong dependency on fossil fuels, rising congestion and increasingly fossil fuel limits.

This chapter develops a vision to 2050 for the EUFT sector by using causal storylines that are backcasted from the long-term future. A causal, storyline is a powerful tool of packaging a complex set of events and relationships into something that is cognitively manageable and therefore memorable (GBN 2011, p. 7). A set of plausible storylines are developed that break the trends from the origins of the stories about the future. The storylines that make up the views of stakeholders feed the backcasting approach of the vision for Freight.

The structure of the chapter is as follows: Sect. 3.2 reviews the literature, Sect. 3.3 introduces the method of strategic conversation and visioning, Sect. 3.4 shows the case study of freight transport and Sect. 3.5 reports the results while Sect. 3.6 contains the conclusion.

3.2 Literature Review

The literature on transport engineering and economics deals with transportation planning mostly as a rational process based on the formulation and comparison of alternative options (Cascetta et al. 2015). The rational process has been part of the traditional method which relies on 'predict and provide' policy practices usually using mechanistic models of reality through econometric analysis of transport demand. 'Predict and provide' refers to setting out to accommodate an expansion of traffic through a large programme of road building, however additional road capacity generates additional traffic. This practice has dominated British and European transport planning for decades. The tools of forecasting freight traffic rely on econometric analysis through statistical techniques because these enable predict and provide approaches for freight transport decision-making, however, forecasters generally do a poor job of estimating demand for transportation infrastructure projects (Flyvberg et al. 2006).

---

[1] Freight transport includes: road freight, rail freight and inland waterways freight. We exclude airfreight from this study.

An alternative to that traditional method is the formulation of policy recommendations by applying the backcasting exercise using stakeholders' knowledge of the sector to take action and to participate in the process. The backbone of backcasting research is to spur collaboration across the entire supply chain to arrive at better decisions.

There are few studies on the freight sector that are based on the backcasting techniques, using stakeholder led tools, within the context of sustainability. Stakeholders of the freight sector are a key input for applying group reasoning techniques (Raiffa et al. 2002; Yearwood and Stranieri 2011). The use of stakeholders can be justified because such methods can help design the research agenda by designing hypothesis, collecting data and setting goals for the sector.

There is a large number of backcasting studies which do not use stakeholder input and yet these studies have elicited solutions for future policy guidance (Akerman and Hojer 2006; Fujino et al. 2008; Van Wee & Geurs 2004). The latter investigates the Dutch freight transport system to the year 2030.

Backcasting studies that rely on stakeholder collaboration for the freight transport and logistics sector are few. Two studies stand out in the field of freight transport and logistics. In one study, a wide range of stakeholders are used to build scenarios and rank 25 policy measures (Hans et al. 2012). The scenarios assess clean vehicles and generate two policy scenarios. The scenarios also examine economic instruments (i.e. congestion charge) and technical ones for the case of passenger vehicles. That study's vision is based on the views of a small group of stakeholders; unlike them, our study considers the whole range of stakeholders from the entire supply chain (See Tables 3.3, 3.4, 3.5 and 3.6 in Sect. 3.4.3).

Pioneering studies of freight transport that use stakeholder input, in decision-making, through the MACMA (multi-actor multi-criteria analysis) method are few. An example of this is a second study by Macharis et al. (2010) who developed a multi-stakeholder analysis for logistics to identify transport policy measures. An application of MACMA for evaluation of transport projects (Macharis et al. 2010) shows the strength of the method. Similarly MACMA analysis is developed for the Dutch freight transport and logistics sectors bearing in mind that traditional logistics methods embedded in operations research, inventory control, production scheduling and transport routing are also valuable (Van Duin 2012). An Australian study (Hensher and Golob 1999) relies on stakeholder input for decision-making on Freight organization, however, the study uses only stakeholders from that nation leaving out the weight of non-Australian stakeholders in shaping decisions for freight.

Another study by Gonzales Feliu et al. (2013) concludes "…collaborative logistics and freight transport systems appear to be interesting strategies but each involved decision maker can diverge from the others on the best form of collaboration." One reason for this divergence in action is that decision makers usually belong to different interest groups. (In the case of EUFT this can be seen in Tables 3.3, 3.4, 3.5, and 3.6: there are interest groups from several fields of transport policy and the private sector. See Sect. 3.4.3)

Large parts of the literature have focused on discovering a set of megatrends and driving forces to accompany the vision to 2050. These megatrends are selected through the involvement of stakeholders. But there is no consensus in the freight transport or logistics research field on the key driving forces of change in the sector.

The vision proposed in this study considers the limits to planetary growth. The vision for EUFT is an attempt to what De Jounevenel (1967) (cited in Bradfield et al. 2005, p. 802) identifies: 'to avoid a particular view of the future held by small political groups which determined how the future of the nation unfolded'. The vision proposed reflects the collective wishes (society, firms and governments) for the EUFT system in 2050. The vision is facilitated by using the strategic conversation tool. Strategic conversation uses many mechanisms: meetings, decision-making processes, the planning system, mandatory submissions and documentation (Van Der Heijden 2010). The vision is an image of the future of freight transport to 2050 based partly on the backcasting developed in the FREIGHTVISION (henceforth FVision) project (FVision 2009, 2010a, b; Helmreich and Keller 2011).

The vision for EUFT is developed using a backcasting exercise. The method improves future thinking of freight transport organizations (cargo firms, government ministries, freight forwarders and integrators). Quantitative analyses of the latter are dealt with elsewhere (Mattila and Antikainen 2011).

The backcasting method needs the identification of a particular (closed) future and the tracing of pathways of progress and implementation back to the present (Sharad and Banister 2008). Backcasting is concerned with how desirable a future is rather than with how likely a future can be (Robinson 1982; Lovins 1976; ETAG 2008). Backcasting requires a set of targets that are based on social and ethical views. Backcasting is distinct from traditional scenarios (open futures) and assesses the means by which the closed future or the vision can be achieved. Backcasts' exercises include: (1) specific timelines for action plans or policy innovations and (2) expert feedback through workshops that are used to develop backcasting pathways.

In the freight transport field, seven previous works used the backcasting method for analysis at the single country level. Important backcasting exercises to build images of a future London transport system are developed in Hickman and Banister (2007) and in VIBAT (2008). Other large-scale vision studies include the ETAG project (2008), U.S. Transportation Research Board (2013), and the UK's (CILT 2011). A study for Japan (Fujino et al. 2008) develops two scenarios for freight transport and assumes changes in efficient reciprocating engine vehicle; it also expects diffusion of advanced vehicle technology: hybrid engine, bio-alcohol, electric, plug-in, natural gas and fuel-cell. Further measures of that study include: weight reduction of vehicle, friction and drag reduction in vehicle, efficient railway, efficient ship and efficient airplane. Four additional measures are also considered: intelligent traffic systems (ITS), real-time and security traffic system, supply-chain management and virtual communication systems. The backcasting method has also been widely applied for understanding the future of the passenger sector (Miola 2008) but much less so for the understanding the megatrends of freight transport flows.

The vision for EUFT developed here allows the entire sector to manage more effectively the threats, or improve the options, posed by the six key megatrends; these are discussed in Sect. 3.4.

Five weaknesses can be found in the studies just described, first is the assumption that all scenarios are equally plausible, but scenarios in practice are more or less plausible than others. The second issue is that no actions plans are produced by most of those studies. Third, those studies fail to select a single vision or desirable scenario (future) from an environmental, societal and a corporate lens. Fourth, critics question the effectiveness of analysis which ignores the stakeholders from multiple countries. Lastly, critics question the suitability of open scenarios approaches to provide solutions: in the future the Freight sector is likely to face limits to its own expansion alongside the ecological and resource limits that are likely to affect the wider economy and trade activities. Restricting trade volumes will clearly affect freight transport flows as fewer commodities would need to be moved globally. Transport planners also question the assumption of a positive link between economic growth and freight transport. In Jackson et al. (2014), the limits to economic growth can be traced in terms of ecological limits of the global economy. These limits require a set of targets to reach the desirable future for freight transport flows.

In addition, none of the cited studies include the strategic conversation method to build a vision. Strategic conversation involves developing a common understanding of the future (Frommelt 2008) through open conversation. The vision developed here is the counterpoint to scenarios for coping with turbulence and uncertainty (Wack 1985; Van Der Heijden 2010). This vision is based on the concept of strategic vision (Wack 1985) that leads stakeholders to action.

In all futures and backcasting writers, from Robinson (1982) to Hickman and Banister (2007), Hickman et al. (2011) and from Lovins (1976) to Wangel (2011), no commonality in methodology emerged. Lovins (1976) developed a vision for national energy targets but did not identify the key uncertainties in reaching the vision.[2] Neither did Hans et al. (2012) when writing their vision.

One of the advantages of the backcasting technique applied here is its adaptability for different situations under different circumstances using a variety of viewpoints. This chapter contributes to studies of EUFT policy and to studies of futures by three accounts: by acquiring legitimacy for policy actions, by creating a vision and by finding a common agreement among stakeholders. This is the first study that develops a vision for EUFT to 2050.

Theory on transport planning needs to be further developed that incorporates new perspectives from the freight transport systems and public logistics fields (of Government and wider publics). Traditional work has tended to be excessively focused on econometric analysis for decision-making and for forecasting as already mentioned above.

---

[2] One key difference, with other studies, is the time length of previous studies with the Shell scenarios (open futures) taking the longest amongst the most well-known ones. This is a much longer period than that taken by the FVision workshops.

New perspectives should be based on open dialogue among stakeholders to arrive to better decision-making on freight transport planning. There are four ways forward to enrich the theory of decision-making on freight transport. First, backcasting studies should draw from behavioural economic principles (Kahneman and Tversky 1979; Taleb 2007). Second the strategic management literature (Ramirez et al. 2010) has much too teach in this regard i.e. how to deal with uncertainty in decision-making for planning freight transport investment. Third, too often decision-making in freight transport planning has depended on 'optimism bias' which has led to infrastructure expansion, more roads, bridges to reduce traffic without reducing urban congestion; fourth theory should develop surrounding connections between robotization and mobile applications.

3.3   Methodology

This section discusses the theoretical literature and empirical explanations regarding the future for freight. The section introduces the vision concept as the ideal desired future.

3.3.1   Theory

The section describes the backcasting and its theory, its assumptions and how these can be violated under certain conditions. Figure + encapsulates the theoretical concept of our vision exercise and how it differs from alternative futures.

The vision is a preferred future which is described in Sect. 3.5. The vision implies that the concrete actions should be taken to move as close as possible to the desired future (Sect. 3.5); however, in the absence of the action plans through a mix of policies, the 'probable future' will emerge as in BaU future. Past events and forces shaping the present, depicted in Fig. 3.1, can include megatrends (discussed in Sect. 3.4). A vision can be developed around a target and this can be the basis for the backcasting exercise (Wangle 2011) which is a normative approach.

In this study, we introduce a further element to backcasting methods: we use the strategic conversation method to retrieve information from the stakeholders. The backcasting method assumes that a single vision can be formed under the general agreement of the stakeholders. This assumption cannot always hold and achieving agreement on the vision is not always possible or meaningful; for this reason, we included a wide range of stakeholders in the exercise. Figure 3.1 shows how past events and forces shape today's context of Freight. A common problem with visioning is that it is difficult for stakeholders to adopt a common definition of the present context facing the EUFT sector i.e. the competitive context of the freight market, the macro-economy etc. A second problem is that the vision exercise does not explore in great depth the large uncertainties as the year 2050 target is approached.

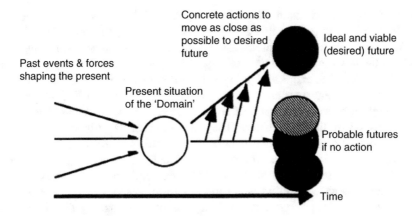

Fig. 3.1 Ideal future: the vision for freight. Adapted from Ramirez (1993)

The backcasting method combined with the foresight approach is used to generate the vision for 2050 Freight. The foresight methods vary widely but in the EUFT vision three elements are used: expert based techniques, interaction based techniques (through strategic conversation) and evidence based techniques. Expertise, creativity, interaction and evidence make up the diamond of foresight techniques (Popper 2008). Foresight uses qualitative and quantitative elements to assess the future (Miles et al. 2008).

The purpose of visions and scenarios in transport studies differs from user to user. Corporations will use the latter to design corporate strategy and risk (Schwartz 1996) and policymakers will use the former to encourage societal change (Helmreich and Keller 2011).

3.4   Case Study: European Freight Transport

In this section, we apply the above described backcasting method to a case study on the freight transport developments.

3.4.1   Preparatory Steps: Building the Vision

Following the assessment of the contextual environment and of the contractual one, targets are set for a sustainable Freight. The first of these two environments is defined as the wider economy and the second as the actions of actors (firms and governments). The vision or scenario writer must make decisions about where to draw the boundary between the contextual and transactional environments before they develop scenarios (Lang and Allen 2008). These targets (backcasts), which support the vision, are expected to diverge to show the relative policy implications of alternative energy futures (Robinson 1982, p. 38); in this case, futures for EUFT. The 2050 targets for EUFT are derived by analysis of the megatrends.

All workshops were held in English, which may bias the results since linguistic ability affects the clarity of storylines generated by the stakeholders (see list in Tables 3.6, 3.7, 3.8, and 3.9). During the workshops stakeholders had to rethink their hypothesis and their assumptions on the evolution of the Freight sector to 2050. Within the workshop multiple layers were set up; megatrends and storylines were selected by the stakeholders. They also selected the driving forces of the Freight transport system. Using data on forces and megatrends, the vision was developed for each decade starting in 2050. Finally a backcasting exercise was undertaken to explain how the targets would be reached. The starting year was 2050. The vision was tested at every workshop to allow stakeholders to experience the future anew. The final step involved defining a set of policy packages; this was done by the entire group of stakeholders.

A policy package is a combination of individual policy measures, aimed at addressing one or more policy goals. The package is created in order to improve the impacts of individual policy measures, minimize possible side effects and/or facilitate measures' implementation and acceptability (Optic 2010). To examine the effect of the measures in combination the contribution of each of the 18 portfolios was measured by using metric for the 'contribution to growth' in the 'Y' axis and their absolute contribution in the 'X' axis in a 2×2 matrix. Stakeholders were questioned regarding the strengths of the actions of each policy portfolio.

3.4.2   Operational steps

We discuss the operational steps of the case study: this involves the selection of (1) the megatrends, (2) defining the state of the freight sector and (3) understanding the type of actor. The key megatrends shape the likely development, and probable shifts in how Freight is moved by 2050. Megatrends are key to explaining the preferred future for Freight. (A megatrend is defined as a great force in societal development [that is state, market and civil society] that is likely to affect the future in all areas in the next four decades.) Megatrends are the forces that define our present and future worlds and the interaction between these is as important as each individual megatrend (CIFS 2012).

During the workshops, six core megatrends emerged from our strategic conversation with stakeholders: rising CO_2 emissions requiring deep cuts in emissions of European economies, rising fossil fuel dependency, increasing congestion for all modes of transport, growing demand for services of Freight, longer supply chains and the adoption of information technology. Some of these megatrends have dominated the Freight growth in previous decades but others are rapidly emerging as the key ones. Some megatrends are currently underway and their combined effects have yet to be seen in the Freight system. This includes new information technologies: big data, online shopping, and trading, GPS, cloud computing and intelligent labelling.

Table 3.1 Freight Transport performance

Mode	1970 (a)	2010 (b)	2012	2015	(BaU) 2050 (c)
Road Freight Billion t-km	443	1586	1621	1722	2525
Rail Freight Billion t-km	446	390	405	418	790
IWW Freight Billion t-km	108	147	149	148	315

Source: Eurostat (2012a, b, 2014, 2016c)

Table 3.1 shows the most recently available data for the EUFT sector. Table 3.1 shows the BaU (business as usual) future of Freight within the 27 member states. As the table shows, Freight is dominated by the road mode in the European Union (Table 3.1) and this mode is likely to continue to do so under the BaU narrative by 2050. Table 3.1 shows greenhouse gas data converted to CO_2 (Carbon dioxide) equivalents. The share of rail will almost double within the whole freight sector and the share of the IWW mode will treble by 2050. The BaU estimates for freight activity are computed by the Transtools II model; this is a European wide model and it includes 1441 zones of Europe. It includes a passenger demand model and a freight model. The latter model is subdivided into the trade, the mode choice and the logistics models.

Within the 1970–2015 period, Freight performance (using t-km; for EU-24 member states for all modes) has more than quadrupled with a growth rate of 5.5% annually (ITF 2008; EU Commission 2012, 2017). Road freight has seen the largest growth rates in the entire period (3.4% annually), and this mode has been the largest contributor to the rapid growth of the total freight moved (Table 3.1).

Table 3.1 reveals that in the 1970s, low carbon freight (inland waterways and rail modes) had a much larger role in total freight movements than is the case for 2012. The share of rail freight of total Freight was at a par in 1970 with road freight but the latter sector adopted better and cheaper logistics practices, which more easily improved its efficiency. Rail freight was approximately 446 billion tkm (Perkins 2012) in the early 1970s but declined in the following decades because of: (a) the use of freight revenues to cover passenger costs; (b) rail tariff controls and (c) higher labour costs than road freight (Perkins 2012). Rail cargo also declined because of long transfer times.

The changes in modal shares, in 1970–2015, of total Freight (Table 3.1) have two implications: (1) the road freight mode has acquired more influence in transport policymaking than the railways mode has managed to do; (2) a new structure of the business (stakeholders) has emerged, favouring the road mode until 2050.

In the last four decades, the growth of Freight has been facilitated by the growth of world trade, the concentration of production in fewer sites to reap economies of scale, delocalization, and Just-in-Time practices (Eurostat 2012a, b). To these factors one can add energy prices, cheaper information technology, the rise in collaboration across borders between consumer and producers (Friedman 2008), and

Table 3.2 Greenhouse gas emissions

	GHG emissions (Mt-CO_2 eq.)	GHG emissions (Mt-CO_2 eq.)	GHG emissions (Mt-CO_2 eq.)	GHG emissions (Mt-CO_2 eq.)	GHG emissions (Mt-CO_2 eq.)
	2005	2010	2012	2015	2050 (BaU)
Road Freight	190	195	210.8	222.9	276
Rail Freight	10.4	7.3	7.1	6.4	19.7
IWW freight	4.7	19.2	17.2	16.4	11.02

Eurostat (2012a, b, 2014) and Helmreich and Keller (2011); (EU 27 member states). GHG direct and indirect emissions are included

less acknowledged, the adoption of containers in trucks, rail and ships allowing cuts in labour costs of the Freight sector (Levinson 2006). The establishment of the single European market and the opening of China's economy also explain the growth of the Freight sector. In recent years growth of Freight has been seen in the new member states of Eastern Europe rather than in the core members.

The above factors improved road freight performance relative to gross domestic product (GDP). The evidence shows, however, that the EUFT growth decouples slowly from GDP growth for the EU-27 member states since 1990 up to 2010. The 2010–2012 recovery in trade within the EU and in the extra EU trade, as well as the higher GDP growth rate raises the growth of EUFT in 2010–2015.

The rapid growth of the EUFT sector since 1970–2015 has translated into higher (CO_2) emissions (Table 3.2). Freight (EU-27 member states) accounts for roughly one quarter of total carbon dioxide by the entire transport sector in 2012 in the EU.[3] In the future, Freight activity will need to adapt to a world economy with CO_2 constraints and the largest increase in Freight will come from the road freight mode by 2050 (Table 3.1) with total EUFT emissions rising from around 223 Mt-CO_2 eq. (in 2015) (Table 3.1, all modes) to 276 Mt-CO_2 eq. between 2015 and 2050 if no action is taken (Helmreich and Keller 2011). These emission pathways of 2010–2050 are idealized in the vision.

In summary, the megatrends surrounding Freight (and the challenges that have been presented above) form the basis to generate the targets for the vision of freight to 2050. To build the vision we first consider the BaU future for Freight, whilst the second path embraces the vision that challenges the BaU view on the future of Freight; this is done by bringing idealistic images into the political and corporate arenas where they could serve as a guiding vision to policymakers by providing a basis for action (Bradfield et al. 2005).

---

[3] Globally the freight transport sector (combining the sectors of the EU-28 Member States and the rest of the world) accounts for roughly 47% of total energy use in 2006 (all modes included) and the share is growing from 38% (Gilbert and Perl 2008). Trucking energy use accounts for one quarter of total energy use of world transport (Exxon-mobil 2013).

FREIGHTVISION Forum

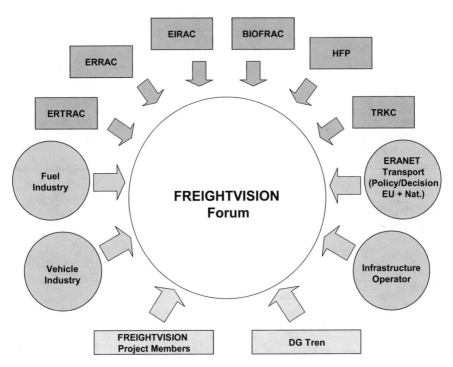

Fig. 3.2 Overview of the contractual environment. (Source the authors based on Helmreich and Keller (2011))

3.4.3 The Role of Actors (Stakeholders) and the Contractual Environment

Figure 3.2 shows the contractual environment.

Each of the actors shown (i.e. the vehicle industry) in the figure will relate differently to the contextual environment. (i.e. the wider economy, oil prices, trade patterns, etc.).

For example on decision-making, stakeholders (from the contractual environment) will react in distinct ways following a diesel price decrease or increase (contextual environment). In addition, the vehicle industry, the fuel industry or the railway companies will react in different ways to changes in the contextual environment. Figure 3.2 describes the type of stakeholders. Figure 3.2 is a snapshot of some of the participants (from the contractual environment) who are more influential in shaping the vision of Freight to 2050 as they hold larger commercial interests in the overall European truck market (Tables 3.3 to 3.6).

The key stakeholders are: EIRAC European intermodal Research Advisory Council; EUTRAC European Transport Advisory Council; BIOFRAC Biofuel

Table 3.3 Stakeholders: corporations

Stakeholder	Interest Group	Meetings	Mode
ADAS Man. Cons.	DG TREN	2	Road
ADEME	UIRR	1	Road, rail
ADIF	UIRR		N.A
AEA Technology	UIRR	1	Road
Alcoa Europe	UIRR		N.A
Alpen-Initiative	UIRR	2	N.A
Alstom	UIRR		Rail
AMRIE—Alliance of Maritime Regional Interests in Europe	DG TREN		Inland waterways
AMRIE—Alliance of Maritime Regional Interests in Europe	UIRR		Inland waterways
APAT	UIRR		Road, rail
ARCESE SPA	UIRR		Road, rail
Arsenal Research	DG TREN	3	N.A
ASECAP	DG TREN		Road
ASSOFERR	UIRR	1	Rail, intermodal
ASTI	UIRR	1	Logistics
ATOC	UIRR	1	Rail
Auchan	DG TREN	1	Road, air
Autoroute Ferroviaire Alpine	UIRR	1	Rail
Autostrade	DG TREN	1	Road
AZP Public agency for rail transport	UIRR	1	Rail
Banque Européenne dInvestissement	UIRR	1	All modes
Banverket	UIRR	1	N.A
BASt	DG TREN	1	N.A
BASt	DG TREN	1	N.A
B-Cargo	UIRR	1	Rail
B-Cargo	UIRR	1	Rail
Belgium	DG TREN	1	N.A
Bertschi AG	UIRR	1	Logistics
BGL Brüssel	UIRR	1	N.A
BIC	UIRR	1	N.A
BMVBS	DG TREN	1	N.A
BMW	DG TREN	1	Road
Bombardier Transportation	UIRR	1	Rail and IWW
Bosch	DG TREN	1	Road
Bosch	DG TREN	1	Road
BSL	UIRR	1	N.A
BTS Kombiwaggon Service	UIRR	1	Rail
Bulgaria	DG TREN	1	N.A
Bundesverband Güterkraftverkehr Logistik Und Entsorgung (Bgl) E.V.	DG TREN		Logistics

Table 3.4 Stakeholders, corporations

Stakeholder	Interest Group	Meetings	Mode
CDV	DG TREN		N.A
CEFIC—European Chemical Industry Council	DG TREN		N.A
Cemat	UIRR		N.A
CEN	UIRR		N.A
Centre de Recherches Routières	UIRR		N.A
CEOC International	UIRR		N.A
CER, Community of European Railways and Infrastructure	DG TREN		Rail
Chambre Régionale De Commerce Et D'Industrie Provence Alpes Côte D'Azur—Corse	DG TREN		N.A
CIDAUT	DG TREN		N.A
CILT—Chartered Institute of Logistics and Transport	DG TREN		Mainly road
CIT	UIRR		N.A
Civil air navigation services organization	DG TREN		Air
Clean Air Action Group	UIRR		N.A
Daimler-Chrysler	DG TREN	2	Road
DB AG	UIRR	2	Rail
DB Deutsche Bahn AG	UIRR		Rail
DB Intermodal Services GmbH	UIRR		Rail
DEKRA	DG TREN	2	N.A
Delphi France SAS	DG TREN		N.A
Deutsche Bahn	DG TREN	3	Rail
Deutsche Post AG, DHL Express	UIRR	1	Rail and Road
Deutsche Post/DHL	DG TREN	1	Rail, road
Deutsches Verkehrsforum	DG TREN		N.A
Energy Solutions Europe	UIRR		N.A
ENTPE Laboratoire Economie des Transports	UIRR		N.A
ETRA I + D	DG TREN		N.A
EUCAR	DG TREN		N.A
EuroCommerce	DG TREN		N.A
EuroCommerce (retail, wholesale and international trade)	DG TREN		N.A
EuroCommerce/IKEA	DG TREN		N.A
Eurofer	UIRR		N.A
HaCon Ingenieurgesellschaft mbH	UIRR	2	N.A
Hugo Häffner Vertrieb Gmbh & Co. Kg	DG TREN		N.A
ICF, Intercontainer	UIRR		N.A
ICSO	UIRR		N.A
Ihk Schwarzwald-Baar-Heuberg	DG TREN		N.A
			N.A

Table 3.5   Stakeholders, corporations

Stakeholder	Interest Group	Meetings	Mode
Ineris	UIRR		
Mazda Motor Logistics Europe	DG TREN		Road
NIKE	DG TREN	1	Road
Procter & Gamble Europe	DG TREN	1	Rail, road, air
Quadrante Servizi S.R.L.	DG TREN		Rail, road
Rail Freight Group	UIRR	2	Rail
Rail Net Europe	UIRR		Rail
Rail4Chem	UIRR		Rail
Region Halland	DG TREN		All three modes
Region Handwerkskammer Reutlingen	DG TREN		N.A
Region Heilbronn-Franken	DG TREN		N.A
Region Landkreis Cloppenburg	DG TREN		N.A
Region Skåne	DG TREN		N.A
Renfe-Operadora (Spanish railways)	DG TREN		Rail
RFF Réseau Ferré de France	UIRR		Rail
Roder	UIRR		Road
RWS	DG TREN		Road
Service Public Fédéral Mobilité Et Transports	DG TREN		Rail, road, air
Service Public Fédéral Mobilité Et Transports	DG TREN		Rail, road, air
SGKV e.V.	UIRR		Logistics (general)
Shortsea Promotion Centre Finland	DG TREN		Rail, road, air
Siemens SA Transportation Systems	UIRR		
Silopor (Empresa De Silos Portuários, S.A)	DG TREN		Shipping
SRA	DG TREN		N.A
T & E	UIRR		Road
T & E	UIRR		Road
Tcl	DG TREN		Rail, road
Teleatlas	DG TREN		Road (navigation and location-based services)
The Netherlands	DG TREN		N.A
TLF, Fédération des entreprises de transport et de logistique de France	DG TREN		Logistics (general)
TLN, Transport en Logistiek Nederland	UIRR	3	Logistics (general)
TNO	DG TREN		Road
Toyota Motor Europe	DG TREN		Road
UPS—United Parcel Service	DG TREN		Logistics (general)

(continued)

Table 3.5   (continued)

Stakeholder	Interest Group	Meetings	Mode
VDE	DG TREN		Road
Verband Für Spedition Und Logistik Der Tschechischen Republik (Ssl)	DG TREN		N.A
Volkswagen	DG TREN	1	Rail, road, air
VOLVO	DG TREN	4	Road
Volvo Logistics	DG TREN	1	Logistics (general)
VTI	DG TREN		
VTT Technical Research Centre of Finland	UIRR, DGTREN		Road
ÖBB Brüssels	UIRR		Rail
AAE Ahaus Alstätter Eisenbahn	UIRR		Logistics (general)

http://www.made-in-germany.biz/en/companies/classified-directory.html

Research Advisory Council; ERTRAC European Road Transport Research Advisory Council; ERRAC European Rail Advisory Council; ERANET European Research Arena; HFP European hydrogen fuel cell technology platform; TRKC Transport Research Knowledge Center and DGTREN (now DGMOVE).

In addition, there were representatives from both the private and the public sectors, the trade unions, the transport ministries of key EU member states, infrastructure operators, industry (fuel, vehicle) and lobbying organizations. Three of the consortium partners belong to federal bodies; these were included to assure a broad mix of experts. There were seven European parliament members and ten officials from the EU. There also were executives from logistics companies, infrastructure operators, cargo owners, and from the vehicle making industry. In total, there were approximately 110 stakeholders at meetings and discussions, belonging to a broad range of fields so obtaining a wide variety of viewpoints.

Tables 3.3, 3.4, 3.5, and 3.6 tabulate the participants, the number of attendees per meeting, numbers of meetings and the mode they are likely to belong to. Some participants are likely to belong to more than one mode i.e. UIRR participants. The stakeholders are subdivided by corporation, transport ministries or local authorities and associations.

As far as the participants are concerned, a conflict emerges between, for example, rail and road freight forwarders, or between national regulators and truck makers (Fig. 3.2). This is reflected in the policy measures enshrined in the vision of Freight. The interaction of these actors (Fig. 3.2) under turbulence in the contextual environment will determine the future growth trajectory of the freight transport sector. Turbulence refers to 'when the whole common shared ground is in motion (Ramirez et al. 2010)' One stakeholder opined on the vision as in '…UNIFE is concerned that the Action Plan focuses excessively on road transport, not least the electrification of cars...' (FVision 2010a, b). The measures emanating from the vision, however, do consider the electrification of both truck fleets and train fleets; both modes play a full part of the vision for Freight.

Table 3.6 Policymakers

Stakeholder	Interest group	Meetings	Mode
BMVIT, Federal Ministry of Transport Innovation and Technology (AT)	DG TREN	2	Rail, road, air and domestic shipping
Austrian Federal Economic Chamber (AT)	DG TREN	1	Industrial classification (Transport and Communications)
Antwerp Port Authority (BE)	UIRR	1	Shipping, road, logistics
Centre Conjoint OCDE de Recherche/ Transports	UIRR		n.a.
European Parliament		2	Rail, road, inland waterways, air and shipping
Ministry Brussels Region	DG TREN		Road, rail
Ministry of Economic Affairs and Communications (BE)	DG TREN	1	Road, air and shipping
Ministry of Economy and Transport (EE)	DG TREN	1	n.a.
Ministry of Enterprise, Energy and Communication (SE)	DG TREN	1	Rail, road, air and shipping
Ministry of Infrastructure (unknown country)	DG TREN	1	Rail, road and shipping
Ministerio de Fomento (ES)		1	Rail, road, air and shipping
Ministry of Transport (NL) (1967–2010) Now: Ministry of Infrastructure and Water Management. Directorate-General for Mobility and Transport (DGB)	DG TREN	3	Rail, road, air and shipping
Ministry of Transport and Communications (FI)	DG TREN	1	Rail, road, air and shipping
Ministry of ecology (FR) Ministry for an Ecological and Solidary Transition (FR)	DG TREN	3	Rail, road, air and shipping
Ministry of Transport and Communications (NO)	DG TREN	1	Rail, road, air and shipping
Ministry of Transport and Communications (FI)	DG TREN	1	Rail, road, air and shipping
Ministry of Transport, Construction and Tourism (LT) Ministry of Transport and Communications (LT)	DG TREN	1	Rail, road, inland waterways, air and shipping
Ministry of Transport, Construction and Urban Development (RO) Ministry of Transport (RO)	DG TREN	1	Rail, road, inland waterways, air and shipping
Ministry of Transport, Posts and Telecommunications (SK) Ministry of Transport and Construction of the Slovak Republic (SK)	DG TREN	1	Rail, road, air and shipping

(continued)

Table 3.6 (continued)

Stakeholder	Interest group	Meetings	Mode
Ministry of Transport (DE) Federal Ministry of Transport and Digital Infrastructure—BMVI (DE)	DG TREN	4	Rail, road, inland waterways, air and shipping
Municipality of Strömsund (SE)	DG TREN	1	n.a.
Direction Des Transports Calais	DG TREN	1	Rail and road
Service Public Fédéral Mobilité Et Transports (BE)	DG TREN	1	Rail, road, air and shipping
EC DG-INFSO Now DG CONNECT	DG TREN	3	Rail, road, inland waterways, air and shipping
EC DG-TREN	DG TREN	3	Rail, road, inland waterways, air and shipping
EC European Commission RTD	DG TREN	2	Road
EESC—European Economic and Social Committee	DG TREN	1	Rail, road, air and shipping
ECA European cockpit association	DG TREN	1	Air
Cyprus Port Authority (CY)	DG TREN	1	Road and shipping
Service Public Fédéral Mobilité Et Transports (BE)	DG TREN	1	Rail, road, air and shipping
Service Public Fédéral Mobilité Et Transports (BE)	DG TREN	1	Rail, road, air and shipping
Permanent Representation of Romania to the EU (BE)	DG TREN	1	n.a.
Polish Ministry of Transport (PL)	DG TREN	1	n.a.
Port of Rotterdam (NL)	DG TREN	1	Rail, road and shipping
Ports Directorate—Malta, Now Ports and Yachting Directorate (MT)		1	Shipping
Department for Transport (UK)	DG TREN	2	Rail, road and shipping
UK Permanent Representation to the EU	DG TREN	1	Road

Table 3.7 The storylines collected from the strategic conversations exercises

GHG	Fossil fuel dependency	Congestion
Stable population	Peak oil will be reached in the near future (2015).	Infrastructure development for co-modality
Economic growth	There will be an increasing gap between goods demand and production.	Passenger modes will increase at the same rate as that of Freight.
No more tonnes lifted of cargo	Energy efficiency will reduce this gap.	Shipping behaviour will change (collaboration amongst freight forwarders).
Global division of labour peaks.	Instability of fuel price; some stabilization of fuel price will be needed.	Mobility increases.
Goods packaging becomes lighter.	Price of energy will rise nonlinearly.	Policy driven shift from to road to rail and ports.
Increase in local production and less intercontinental transport	Not all energy sources will be accepted by society (i.e. oil sands in Canada).	Passenger priority.
Policy actions fail to achieve the 50% reduction in GHG by 2050	Fossil fuels consumption will be phased out when better alternatives become available but the lack of these solutions will need to be found sooner.	Lack of financing.

Source: The authors and FVision (2009)

It was noted that the business and Government relationship was affected by all of the actors shown in Fig. 3.2. These groups of actors, part of the business elite, will informally have relationships that affect how decisions are made that impact on policy measures of freight transport. These groups will help to determine the type of policy that is chosen for the vision.

3.5   Case of Practical Application: Quantifying the Vision

This section presents the results concerning the storylines, the data gathering, the selection of the indicators, the freight transport targets, the implementation of the vision and the driving forces. All of these information is selected by the stakeholders (Tables 3.3, 3.4, 3.5, and 3.6). Table 3.7 describes the story lines on economic growth, on population surrounding the three dimensions of sustainability: GHG, fossil fuels and congestion. Finally we describe the action plan which differs from the horizon 2020 targets (Table 3.8). To develop the vision for Freight to 2050, an imaginary action plan for the sector is recreated into the future by allowing stakeholders (Fig. 3.2) to experience the future through visioning.

Table 3.8 Target for reducing GHG and fossil fuel share (road, rail and IWW)

	2011	2050
Diesel engine (%) efficiency	42%	61%
Targets for rail energy use (MJ per t-km moved).	0.12	0.09 (Gain: −20%)
–	0.10 (MJ)	0.08 (Gain: −20%)
Low carbon electricity	0.12 (kg)	0.015 (kg)
Biofuels (CO$_2$ eq. in kg/MJ)	0.10 (MJ)	0.12 (MJ)
Biofuels blending (in %)	2	33
Efficient use of vehicles. Includes: empty runs, load factors, driver behaviour, etc.	n.a.	50%
Electric energy in road transport (fraction of electricity in road transport energy use).	n.a.	25%
Modal split (billion t-km) Road	1635	2045
Rail	415	790
IWW	135	315
Truck weight dimensions (% of t-km transported by larger trucks: 40 t load, Gigaliner)		10
Electrification of the Railways Diesel	64%	80%
Electricity	36%	20%

Source: Helmreich and Keller (2011)

3.5.1   The Storylines of Stakeholders

Before developing a vision based on the criteria of three dimensions, we present a list of key storylines that are instrumental in developing the vision and identifying the key drivers. The storylines describe a very heterogeneous world for Freight from the viewpoint of the various actors. Table 3.7 shows examples of storylines. The storylines guide the backcasting exercise, its goals and the steps needed to arrive to the desired future. For example changes in the new global division of labour can lead to a new fragmentation of industry and thus to changes in the distribution of manufacturing establishments across the globe. These changes will affect freight transport flows and modal shares (Fig. 3.3). See also Table 3.9 in Sect. 3.5.4.

3.5.2   Setting the Targets

The storylines provided the backbone for setting the targets. Setting the targets for future freight and how this sector diverges from its BaU future is essential for building the vision for freight. The targets of 'a desired future' and the action plans were taken from the storylines of the stakeholders and were thus considered in the vision of Freight.

The action plans aim to close the gap between the target of the vision described in this article and the BaU future. The action plan includes many dimensions that allow achieving a target for 2050, in this case for fossil fuel dependency (Table 3.3). The action plan can support a transition to decouple freight transport from GDP growth (McKinnon 2007).

Some of the targets for every criteria (i.e. CO_2 emissions reduction) were defined by the stakeholders. One difficulty the stakeholders faced was deciding: (1) the 2050 target (e.g. CO_2 emissions by 2050); (2) the baseline year for both cutting CO_2 emissions and fossil fuel use and (3) the timing of mitigation efforts to reduce congestion. Some stakeholders favoured immediate cuts in CO_2 emissions Freight; others favoured postponing this date (2020). The railway related stakeholders proposed early CO_2 emission cuts whilst other stakeholders wanted a delay in emissions cuts; some stakeholders were neutral about the starting date for emission cuts. Vehicle-

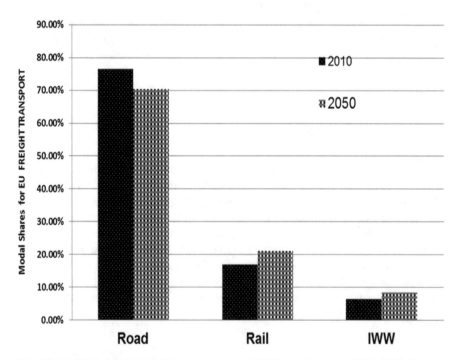

Fig. 3.3 Modal % shares for freight transport in the EU 27 member states 2010: 2050. (Source: FVision project (2010a, b))

related stakeholders favoured slow reductions for their sector. One reason for this discrepancy is that the railway mode is less fuel intensive than the road mode is.

The targets of the vision require a number of future changes in, for example, truck engine technology (Table 3.8) as well as transport policy. The vision is fed by the views and beliefs of stakeholders (these are listed in Tables 3.2). For example, the vision for GHG reductions: stakeholders combined the effect of modal split, engine efficiency and fuel switching to achieve the 80% reduction in GHG emissions of the entire freight sector by 2050. These targets (Table 3.8) represented achievable energy savings after adopting technical improvements; however no investment cost considerations have been made. The targets are also included in the action plans (see Sect. 3.5.7). The target for diesel engine efficiency is based on U.S. DOE (2006) which assumes an increase in efficiency to 60% is feasible by 2050. Another key target is efficient management of fleets i.e. reducing empty running of trucks. A big gain for truck energy efficiency can also be achieved by increasing truck size, which according to stakeholder opinion can take 10% of total freight moved in the EU by 2050 (Helmreich and Keller 2011). Further gains in energy efficiency are possible through (1) higher load factors of trucks by 50% and (2) electrification of trucks and biofuels processing improvements (Telias et al. 2009; FVision 2010a, b). These two latter measures should reduce the GHG emissions. As for the rail mode, it is foreseen that further electrification is possible as well as further dieselization of the rolling stock. The IWW mode will need to achieve a further decline in the share of freight moved by this mode.

The rest of the targets are either based on stakeholder views or on desk research (Table 3.8). These targets were compiled following discussions with stakeholders i.e. truck makers, freight forwarders, distributors and transport operators. The targets were complemented by using published literatures on the theme.

The assumptions (Table 3.8) were tested at the interviews and conversations at the workshops.

3.5.3   Implementing the Vision

A visioning process is implemented to achieve stakeholders' support for the action plan. For this purpose the four forums allow for the active involvement of representatives from various fields (Fig. 3.2). Figure 3.3 shows the modal shares to 2050 that are assumed in the 'BaU' future.

The vision exercise for freight transport to 2050 sheds light on the experience of the users of the vision (logistics companies, freight forwarders, policymakers, stakeholders) as much as for the expectator (the actual researcher) who builds the future. The storylines that are generated at the forums also generate a series of key drivers for the vision (Sect. 3.2). The road freight share declines but unless measures are taken it will continue to be the main transport mode to 2050 (Fig. 3.3).

Since the 1970s until 2005 there has been a sharp decline in the share of rail freight (0.44% annual decline) in the European Union. The pipeline mode, with an

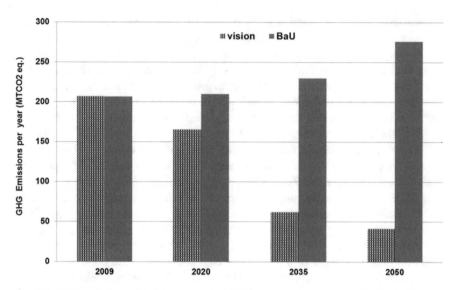

Fig. 3.4 GHG emissions of freight transport of 27 European member states: idealized pathways to 2050. The chequered bars represent the vision for minus 80% GHG. (Source: the authors and The FVision (2010a, b))

annual growth of 4.5% per year, has reduced the amount of road and rail freight over the same period. Figure 3.4 describes the CO_2 emission path of the vision.

The outcome of the vision is the *80% reduction in CO_2* emissions by 2050 and any delay in *CO_2* reduction is costly and dangerous (Stern 2007). The gap between the BaU future and the vision (the preferred future) is closed by means of an action plan as we explain below.

3.5.4   GHG Emission Reduction: The Vision

Figure 3.4 above also describes the quantitative targets of the vision for decarbon-izing the freight transport sector of the EU-27 member states for each target year. As shown in Table 3.8, the vision rests on technological developments (improved engine efficiency and aerodynamics), policy innovations (national road charging, fuel taxes, including the EUFT sector in the emissions trading scheme); (see also Sect. 3.5.5 below). The vision adopts the precautionary principle in CO_2 mitigation as well as encouraging early action on *CO_2* mitigation, since it is cost effective to contain the cumulative increase in GHG emissions (Stern 2007). [4]

---

[4] The precautionary principle argues that it is better to act now rather than later since the conse-quences of not acting outweigh those in its absence.

Table 3.9 Examples of storylines to 2050 extracted from the stakeholders to 2050

Technological change continues, leading to fuel efficiency improvements for trucks.	Change in lifestyle (lower level of goods consumption).
Early action in CO_2 mitigation and fossil fuel dependency continue.	Oil imports dependency declines to 50% of today's level.
High learning rates of new technology allowing market diffusion of fuel efficient trucks.	Fuel efficiency standards on new trucks are imposed.
Cost decline of engine efficient technologies are linked to past investments.	Competition among the modes favours the train mode.
CO_2 abatement costs are low; EU is forced to cut emissions regardless of international cost effective solutions. EUFT emissions cuts are treated separately from other sectors.	The container is handled easily across modes which have become larger, allowing rail to compete freely with trucks.
CO_2 abatement costs fall as learning rates improve under mass scale production of energy efficient vehicles technologies and under renewable energy.	Fuel switching favouring natural gas and biofuels.
Modal shift favouring rail and shipping sectors.	Constant or zero growth of transport volumes of freight transport.

Source: FVision (2010a, b) and Authors

The vision's quantitative targets are revised constantly throughout the vision exercise which is based on storylines (Table 3.9). Three points of disagreement, however, arise: (1) the lack of flexibility of the vision; (2) the need to adopt an open future: a positive scenario rather than a closed one (3) the need to consider a high level of uncertainty regarding these quantitative goals to 2050 (Fig. 3.4).[5] The stakeholders identified the key drivers that can set *off the megatrends.*

3.5.5   Driving Forces and the Business Elite

The driving forces as well as the storylines, described below emanate from the strategic conversations held in various meetings throughout the workshops. The strategic conversation process generates storytelling, giving a rich source for selecting the key driving forces of the future of the Freight sector. These driving forces spring from the discussions on megatrends in Sect. 3.4.2 and shape the images within the vision. Examples of the storylines that were collected from the stakeholders are listed in Table 3.9; the storylines complement the vision and enable stakeholders to identify the needed action plans.

---

[5] The terminology used at the workshops poses problems when the quantitative targets are presented. A few stakeholders, who were invited to participate in building the vision to 2050, were reported to be confused when using the concept of vision and scenario. This proves to be a useful learning outcome as it becomes obvious that many stakeholders are not familiar with visioning and backcasting exercises and they have to be constantly reminded of what a vision involves at each meeting.

3.5.6   The Goals for the Action Plan

This section describes the targets that produced the vision for Europe's future Freight; these targets are achievable through the action plans. For the FVision workshops, data inputs for the three criteria (CO_2 emissions of FT, fossil fuel dependency and congestion) are described in: Mattila and Antikainen (2011), and in Anders et al. (2009).

The backcasting and the vision work are complemented by the main policy targets described in:

- 20–20–20 energy/climate goals proposed by the European Commission (EC White Paper 2001, 2009);
- Stern Report (Stern 2007, 2009);
- Intergovernmental Panel of Climate Change Report (IPCC 2007; Kahn et al. 2007);
- Kyoto Protocol targets on GHG and ongoing negotiations on GHG mitigation;
- 2001 EC White Paper and mid-term review of the EC White Paper (EC 2001, 2009, 2011);
- The 2007 Freight Transport Logistics Action Plan (EC 2007a, b comm.);
- The EU transport white paper proposes 60% reduction in CO_2 (EC 2011, 2014);
- A European strategy for low emissions mobility (EC 2016)

These targets and their assumptions were formulated for each key megatrend (Sect. 3.4.2). The same was done for each action plan and the vision. The latter, however, is not entirely compatible with the Transport White Paper (EC 2011, 2001). The targets are compatible with the EC (2016) policy target regarding low carbon transport. The policy plans enshrined in the EC (2007a, b) logistics action plan are broader and cannot be easily translated into transport volume targets.

In contrast to the EC (2001) strategy on trucks, the action plans differ in four ways. First the vision recommends strong gains in truck fuel efficiency by adopting ambitious fuel economy standards, improved load factors and electrification of road transport so that total truck emissions fall by 2050 significantly. Second, the vision also recommends higher transport costs through phasing in congestion charging and higher fuel taxes. Third, the vision does not take a neutral approach regarding alternative fuels and internal combustion engine: it considers improvements in both fields. Fourth, the EC foresees a 60% decline in carbon emissions by 2050 for the entire transport sector (passenger and freight) and so it does not expect specific cuts in CO_2 emissions for the freight sector.

3.5.7   Action Plan for Freight to 2050

The targets discussed above are achieved partly by policy actions that we now describe. The action plans start in 2050 and are implemented backwards to the year 2012 or close to that year. These plans target the challenges that emerge from the vision described in Sects. 3.5.3 and 3.5.4. The action plans are selected by the stakeholders and their deployment over time is tested at meetings. Thirty five policy actions were identified after the development of the vision to 2050, 2035 and 2005, however, the actions are classed according to 12 key areas for actions by 2050 (Table 3.10), as these actions can achieve the greater share of reductions for Freight activity. The actions are designed to achieve our key goals for freight: (1) cuts in emissions of CO_2; (2) congestion and (3) fossil fuel dependency. For example to reduce freight traffic, emissions and fossil energy, four measures will be key: congestion pricing, fuel taxes, the efficient use of vehicles to raise load levels and shifting the modal split. Distance and fuel pricing can reduce average distance hauled.

Table 3.10 Action plans from the vision exercise to 2050

Policy areas	Effective policy actions	Barriers	Supported by
Engine efficiency	Including CO_2 standards (engine level limits) into heavy goods vehicles. Adopt best available technologies and CO_2 labelling	Heterogeneous stock of truck vehicle engines Diffusion lag time Fuel economy target is challenging	ACEA; DB Schenker Rail GmBh and CER
Transport performance	Network optimization cargo owner; e-freight; transport route planning and control	Lack of investment in infrastructure	Various stakeholders
Vehicle energy demand	Clean vehicle technologies I: Aerodynamics and rolling resistance; best available technologies	New directive for maximum weight of HGV needed for aerodynamic changes	Anonymous
Low carbon electricity	CO_2 labelling; taxing fossil fuels	Heterogeneous heavy vehicle markets and engine sizes	ACEA
Electric energy in road transport	Improved batteries; taxation of fossil fuels; investment in road infrastructure	Battery weight and battery costs	Anonymous
Biofuels	Clean vehicle technologies II: biofuels; taxing fossil fuels	Highly uncertain; competes with food production	Stakeholder
Efficient use of vehicles	Transport consolidation and cooperation; training for eco driving; liberalization of cabotage	Competition law may prevent cooperation among firms; driver education implemented differently	Anonymous

(continued)

Table 3.10 (continued)

Policy areas	Effective policy actions	Barriers	Supported by
Truck weight and dimensions	Modifying the rules for HGV's and dimensions; investment in road infrastructure	Additional legislation is needed; no formal proposal has been made for EU cross-border operations	Anonymous
Modal split	ERTMS (European Rail Traffic Management System); intermodal transport; internationalization of external cost	Lack of investment in infrastructure	European Railway Agency
Electrification of rail	Electrification of rail corridors; CO_2 labelling; taxing fossil fuels	Lack of investment in infrastructure; high investment needs	European Railway Agency; UIRR; Volvo
Infrastructure capacity	Investment in intelligent transport; investment in road infrastructure; internalizing externalities of FT activity	Solving product liability; support for automated transport system	Anonymous
Transport pricing	Internalizing externalities of FT activity Congestion charge scheme (road pricing)	Political opposition insome EU members; pricing policies are sole responsibility of nations	Various transport ministries at the EU level; transport and environment

Source: FVision (2010a, b) and authors

3.6 Conclusion

Despite efforts, published in many White Papers by the EC, to reduce fossil energy use, cut emissions and congestion, the sector continues to this day to rely on the internal combustion engine. However, the exercise presented in this chapter shows that backcasting work can guide policymakers. To guide them towards sustainable freight, practitioners should ask three questions. First how far trucks ought to travel within cities and among these by 2050? Second, how much cargo volume can Europe's economy and infrastructure cope with? Third, how much GHG emissions can European societies believe is acceptable?

Similar questions can be raised surrounding the optimal fleet of freight trucks and associated fossil fuel dependency by 2050. These questions can be answered in order to take strategic decisions in transport planning today. Doing the latter requires the vision as presented here. A different growth path of freight can emerge as a reaction to the possible effects of the megatrends. To cope more effectively with the negative effects of the megatrends shortlisted in this chapter, practitioners need to reduce the effects of uncertainty surrounding trade and oil prices. The vision proposed here builds common understanding about the sector's future leading stakeholders to action.

Using the backcasting approach, we gathered views and beliefs of industry players, leading logistics firms, freight transport ones, freight forwarders and transport Ministries from the European Union. This was done for three reasons. The first is to identify the key megatrends. The second is to set targets to 2050 for the freight sector, and the third is to formulate actions that are needed to close the gap between the vision and the current path which is unsustainable. Building the vision requires the strategic conversation technique allowing a variety of viewpoints of the stakeholders.

The backcasting method allows us to propose vision targets and a variety of actions are recommended in twelve policy areas. Among these are energy efficiency measures for moving freight: gains in engine efficiency using fuel efficiency labelling, better transport performance measures by gains in load factors. To reduce freight traffic and average distance hauled, four measures will be key: congestion pricing, fuel taxes, efficient use of vehicles and shifting the modal split. Further actions to minimize emissions include increasing truck weight, enacting CO_2 standards on trucks, infrastructure financing and new infrastructure capacity for intelligent transport. On the rail area, measures for greater electrification of railways also emerge from the vision. To reduce bottlenecks, these measures should be combined with the adoption of speed limits and of biofuel expansion. In short, the measures recommended in the vision are related to the technical, managerial and supply chain domains. Other measures are based on economic and European transport policy.

The vision is an alternative to traditional forecasting techniques, and it depends on setting targets. Unlike the Deutsche post (2012) and the US TRB (2013) studies, this analysis includes a wide array of stakeholders from more than one nation and for many sectors of the freight sector.

The vision proposed in this chapter provides legitimacy for the actions recommended by encouraging corporate and non-corporate partners to consider the limits to economic growth. Our work could be extended to assess further the role of uncertainty by including three or more explorative scenarios rather than including a single desirable future in the analysis of the freight sector to 2050.

Acknowledgements The overall result of this visioning exercise can be found in the book Freightvision published in 2011 (Helmreich and Keller 2011). This methodological chapter is partly based on the work in Helmreich, S. and Keller H. provided the initial concept for the overall vision. This work would have been possible without the FVision team: S. Helmreich, H. Keller, J. Schmiele, T. Mattila, Antikainnen, R., Dueh, J., Meyer-Ruhle, O., Vellay, C., Jorna, R., Zuiver H., Jammerneg, W., Rosic, H., Bauer, G., and Berry John. As well as the many companies' executives that attended our meetings in Belgium. This article acknowledges funding by the EC. The project was funded (Contract n.: TREN/FP7TR/21927-"FREIGHTVISION") Duration: 01.09.2008.

Research assistance for updating Graphs and Tables is gratefully acknowledged by J. Navarro Guevara. Research assistance by Nihan Akyelken (University of Oxford, Transport Studies Unit, School of Geography and Environment) is also acknowledged.

References

Akerman J, Hojer M (2006) How much transport can the climate stand? Sweden on a sustainable path in 2050. Energy Policy 34(14):1944–1957

Anders N, Knaack F, Rommerskirchen S (2009) Socio-demographic and economic mega trends in Europe and in the World—overview over existing forecasts and conclusions for long-term freight transport demand trends in Europe—deliverable 4.1 of FREIGHTVISION—vision and action plans for European Freight Transport until 2050. Funded by the European Commission 7th RTD Programme

Banister D, Anderton K, Bonilla D, Givoni M, Schawnen T (2011) Transportation and the environment. Annu Rev Resour Environ 36:247–270

Bradfield G, Wright Burt G, Cairns G (2005) The origins and evolution of scenarios techniques in long range business planning. Futures 37:795–812

Cascetta E, Carteni A, Pagliara F, Montanino M (2015) A new look at planning and designing transportation systems: a decision-making model based on cognitive rationality, stakeholder engagement and quantitative methods. Transp Policy 38:27–39

CIFS - Copenhagen Institute for Future Studies (2012) Why megatrends matter. http://www.cifs.dk/scripts/artikel.asp?id=1469. Accessed Oct 2013

CILT - UK Chartered Institute of Logistics (2011) Logistics and transport: vision for 2035. Report, pp 1–21. London, UK. https://ciltuk.org.uk/Policy-and-Guidance/Vision-2035. Accessed Oct 2011

De Jouvenel B (1967) The art of conjecture. Basic Books, New York. (cited in Bradfield)

DHL - Deutsche Post (2012) Logistics 2050: a scenario study. http://www.dhl.com/content/dam/Local_Images/g0/aboutus/SpecialInterest/Logistics2050/szenario_study_logistics_2050.pdf. Accessed June 2015

EC (2001) White paper—European transport policy for 2010: time to decide. http://ec.europa.eu/transport/strategies/2001_white_paper_en.html. Accessed May 2013

Environments: Aligning stakeholders for successful development of public/private logistics systems by increased awareness of multi-actor objectives and perceptions. PhD thesis. TRAIL Thesis Series T2012/6, the Netherlands TRAIL Research School

ETAG (2008) The Future of European Long Distance Freight Transport.Commissioned by STOA and carried out by ETAG. Spe-cific Contract No. IP/A/ STOA/FWC-2005-28/SC27. Ref.: Frame-work Contract No. IP/A/STOA/ FWC/2005–28 August 2008. Report prepared by: Jens Schippl, Ida Leisner, Per Kaspersen, Anders Koed Madsen Institute for Technology Assessment and Systems Analysis (ITAS), Forschungszentrum Karlsruhe in the Helmholtz Association and The Danish Board of Technology (DBT)

EU Commission (2007a) Communication from the Commission—Freight Transport Logistics Action Plan. https://eur-lex.europa.eu/legal-content/EN/ALL/?uri=celex:52007DC0607 {SEC(2007) 1320} {SEC(2007) 1321} [COM (2007) 0607] final. Accessed Dec 2014

EU Commission (2007b) DG TREN report European Energy and Transport. Trends to 2030—update 2005. http://ec.europa.eu. Accessed Dec 2014

EU Commission (2009) Communication from the Commission to the European Parliament, the Council, the European Economic and Social Committee and the Committee of the Regions. Towards comprehensive climate change agreement in Copenhagen. https://eur-lex.europa.eu/legal-content/en/TXT/?uri=CELEX%3A52009DC0039. Accessed Apr 2012

EU Commission (2011) Transport White Paper: roadmap to a single European Transport Area—towards a competitive an resource efficient transport system [COM (2011) 144]. https://ec.europa.eu/transport/sites/transport/files/themes/strategies/doc/2011_white_paper/white-paper-illustrated-brochure_en.pdf. Accessed Nov 2015

EU Commission (2012) EU Pocket Book. Transport, Luxembourg. https://ec.europa.eu/transport/facts-fundings/statistics/pocketbook-2017_en. Accessed Feb 2018

EU Commission (2014) Strategy for reducing heavy-duty vehicles fuel consumption and CO2 emissions [COM (2014) 285]. http://ec.europa.eu/clima/policies/transport/vehicles/heavy/docs/swd_2014_160_en.pdf. Accessed May 2015

EU Commission (2016) Communication from the Commission to the European Parliament, the Council, the European Economic and Social Committee and the Committee of the Regions. A European strategy for low-emission mobility [COM (2016) 501]. https://eur-lex.europa.eu/legal-content/en/TXT/?uri=CELEX%3A52016DC0501. Accessed Oct 2017

EU Commission (2017) EU Pocket Book. Transport, Luxembourg. https://ec.europa.eu/transport/facts-fundings/statistics/pocketbook-2017_en. Accessed Feb 2018

Eurostat (2008) Energy consumption and production. News release. https://ec.europa.eu/eurostat/. Accessed Feb 2009

Eurostat (2012a) Transport in figures. Table on GHG emissions in transport-EU-27-by mode (million tonnes eq., in page 122. Statistical Pocketbook, pp 1–133). https://ec.europa.eu/transport/facts-fundings/statistics/pocketbook-2014_en.htm. Accessed Jan 2015

Eurostat (2012b-2014) Statistical Office of the European communities. "Freight Transport Statistics", Luxembourg. https://ec.europa.eu/statistics_explained/index_php/. Accessed Sept 2015

Eurostat (2014) Transport in figures. Table on GHG emissions in transport-EU-27-by mode (million tonnes eq). http://ec.europa.eu/transport/facts-fundings/statistics/pocketbook-2014_en.html. Accessed Jan 2015

Eurostat (2016) Statistical Office of the European communities. "Freight Transport Statistics", Luxembourg. https://ec.europa.eu/statistics_explained/index_php/. Accessed Sept 2017

Exxon-Mobil (2013) The outlook for energy: a view to 2040, pp 1–53

Flyvberg B, Holm B, Buhl KS (2006) Inaccuracy in traffic forecasts. Transp Rev 26(1):1–24

Friedman T (2008) Hot, flat and crowded: why the world needs a green revolution and how we can renew our future. Penguin Books, London

Frommelt O (2008) Strategy, scenarios and strategic conversation: an exploratory study in the European Truck industry. PhD Thesis, Nottingham University, UK

Fujino J, Hibino G, Ehara T et al (2008) Back-casting analysis for 70% emission reduction in japan by 2050. Clim Policy 8:S108–S124

FVision - Freightvision (2009) Management summary III, assessment of measures and action scenario, pp 1–72. Austria tech. Vienna Austria. FREIGHTVISION—vision and action plans for European Freight Transport until 2050. Funded by the European Commission 7th RTD Programme. https://cordis.europa.eu/project/rcn/90307/factsheet/en

FVision - Freightvision (2010a) Recommendation of a vision and action plans for European policy and European key demonstration projects. Report prepared by S. Helmreich, D. Bonillla, and N. Akyelken and J. Duh and L. Weiss. Deliverable 7.1 of FREIGHTVISION—vision and action plans for European Freight Transport until 2050. Funded by the European Commission 7th RTD Programme. https://cordis.europa.eu/project/rcn/90307/factsheet/en

FVision - Freightvision (2010b) Freighvision: final conference. Feedback stakeholder. Brussels, Belgium 23rd February, 2010. https://cordis.europa.eu/project/rcn/90307/factsheet/en

GBN (2011) Scenarios, strategy and the strategic process. Report prepared by K. Van der Heijden for Global Business Network, US, California, pp 1–33. www.gbn.com. Downloaded on the 20th of October 2011

Geurs K, Van Wee B (2004) Backcasting as a tool for sustainable transport policy making: the environmentally sustainable transport study in the Netherlands. Eur J Transp Infrastruct Res 4(1):47–69

Gilbert R, Perl A (2008) Transport revolutions: moving people and freight without oil. Earthscan, London

Gonzales Feliu J, Morana J, Salanova Grau J et al (2013) Design and scenario assessment for collaborative logistics and freight transport systems. Int J Transp Econ:207–224

Hans M, Carolien M, Schrooten L et al (2012) Exploring the transition to a clean vehicle fleet: from a stakeholder views to transport policy implications. Transp Policy 22:70–79

Hao H, Geng Y, Li W et al (2015) Energy consumption and GHG emissions from China's freight transport sector: scenarios through 2050. Energy Policy 85:94–101

Helmreich S, Keller H (eds) (2011) FREIGHTVISION—Sustainable European Freight Transport 2050: forecast, vision and policy recommendation. Springer, Vienna

Hensher D, Golob TF (1999) Searching for policy priorities in the formulation of a freight transport strategy: an analysis of freight industry attitudes. Transp Res Part E Logist Transp Rev 35:241–267

Hickman R, Banister D (2007) Looking over the horizon: transport and reduced CO_2 emissions in the UK. Transp Policy 14(5):377–387

Hickman R, Ashiru O, Banister D (2011) Transitions to low carbon transport futures strategic conversations: from London to Delhi. J Transp Geogr 19:1563–1562

IPCC (2007) Summary for policymakers. In: Solomon S, Qin D, Manning M, Chen Z, Marquis M, Averyt KB, Tignor M, Miller HL (eds) Climate change 2007: the physical science basis. Contribution of working group I to the fourth assessment report of the intergovernmental panel of climate change. Cambridge University Press, Cambridge

ITF International Transport Forum (2008) Trends in the transport sector: 1970–2006 OECD, Paris

Jackson T, Drake B, Victor P et al (2014). Foundations for an ecological macroeconomics: literature review and model development. https://timjackson.org.uk/www-for-europe/. WWW for Europe working paper no. 65. Accessed Feb 2015

Kahn H, Wiener A (1967) The year 2000: a framework for speculation on the next thirty-three years. MacMillan, New York

Kahn R, Kobayashi S, Beuthe M, Gasca J, Greene D, Lee D, Muromachi Y, Newton PJ, Plotkin S, Sperling D, Zhou PJ (2007) In: Metz B, Davidson OR, Bosch PR, Dave R, Mayer LA (eds) Climate change 2007: mitigation, contribution of working group III to the fourth assessment report of the intergovermental panel on climate change. Cambridge University Press, Cambridge

Kahneman D, Tversky A (1979) Prospect theory: an analysis of decision under risk. Econometrica 46(2):171–185

Lang T, Allen L (2008) Reflecting on scenario practice: the contribution of a soft systems perspective. In: Ramirez S, Van Der Heijden (eds) Business planning for turbulent times, 1st edn. Lang Earthscan, London, pp 47–63

Levinson M (2006) The box: how the shipping container made the world smaller and the world economy bigger. Princeton University Press, Oxfordshire

Lovins A (1976) Energy strategy the road not taken. Foreign Aff 55(1):65–96

Macharis C, De Witte A, Turcksin L (2010) The Multi-Actor Multi-Criteria Analysis (MAMCA) application in the Flemish long-term decision making process on mobility and logistics. Transp Policy 17(5):303–311

Mattila T, Antikainen R (2011) Backcasting sustainable freight transport for Europe in 2050. Energy Policy 39(3):1241–1248

McKinnon A (2007) Decoupling of freight transport and economic growth trends in the UK: an exploratory analysis. Transp Rev 27:37–64

Meadows D (1972) The limits to growth. Universe books, New York

Miles I, Cassingena HJ et al (2008) The many faces of foresight. In: Gheorghiou L, Cassingena HJ et al (eds) The handbook of technology foresight: concepts and practice. Primes series on research and innovation policy in Europe, 1st edn. Edward and Elgar, Northampton, pp 1–456

Miola A (2008) Backcasting approach for sustainable mobility. European commission, Joint Research Center institute for environment and sustainability, pp 1–74. JRC and technical reports. http://publications.jrc.ec.europa.eu/repository/bitstream/111111111/7659/1/backcasting%20final%20report.pdf. Accessed Apr 2015

Nuzzolo A, Crisalli A, Comi A (2015) An aggregate transport demand model for import and export flow simulation. Transport 30(1):43–54

Optic (2010) Deliverable 1. Inventory of measures, typology of non-intentional effects and a framework for policy packaging. Downloadable from http://optic.toi.no. Accessed Apr 2015

Perkins S (2012) Seamless transport policy: institutional and regulatory aspects of intermodal coordination. International transport forum, discussion paper 2012-05, pp 1–26, Paris. https://www.itf-oecd.org/seamless-transport-policy-institutional-and-regulatory-aspects-inter-modal-coordination. Accessed June 2014

Popper R (2008) How are foresight methods selected? Foresight 10(6):62–89

Raiffa H, Richardson J, Metcalfe D (2002) Negotiation analysis: the science and art of collaborative decision making. Harvard University Press, London

Ramirez R (1993) Reinventing Italy: methodological challenges. Futures 28(3):241–254

Ramirez R, Selsky JW et al (eds) (2010) Business planning for turbulent times: new methods for applying scenarios. Routledge, London

Robinson JB (1982) Energy backcasting: a proposed method for policy analysis. Energy Policy 10(4):337–344

Schwartz P (1996) The art of the long view: paths to strategic insight for yourself and your company. Business & Economics, New York

Sharad S, Banister D (2008) Breaking the trend: visioning and backcasting for India and Delhi. Downloadable at: www.vibat.org.vibat/india_report_final.pdf. Scoping report, pp 1–79. The Halcrow group in association with Transport Studies Unit, Oxford University. Accessed Jan 2014

Stern N (2007) The economics of climate change: the Stern review. Cambridge University Press, Cambridge

Stern N (2009) A blue print for a safer planet: how to manage climate change and create a new era of progress and prosperity. Bodley Head, London

Taleb N (2007) The black swan: the impact of the highly improbable. Penguin, London

Telias G, Herndler S, Dirnwoeber M (2009) Engine technologies—internal report 3.2 of FREIGHTVISION—vision and action plans for European Freight Transport until 2050. Funded by the European Commission 7th RTD Programme. https://cordis.europa.eu/project/rcn/90307/factsheet/en. Accessed January 2014

TRB (Transportation Research board) (2013) NCHRP report 750: strategic issues facing transportation, vol 1: Scenario planning for freight transportation infrastructure investment. Report prepared by Caplice, C., Pladnis, S., MIT. Research sponsored by the American association of State highways and transport officials in cooperation with the Federal Highway administration, pp 1–149

U.S. DOE (2006) Review 21st Century Truck partnership road map and technical white paper. Doc. no. 21st CTP-003, Washington, DC, U.S

Van Der Heijden K (2010) Scenarios: the art of strategic conversations. Wiley, Chichester

Van Duin, R. (2012) Logistics concept development in multi-actor

Van Wee B, Geurs K (2004) Backcasting as a tool for sustainable transport policy making: the environmentally sustainable transport study in the Netherlands. Eur J Transp Infrastruct Res 4(1):47–69

VIBAT (2008) Visioning and backcasting for transport. http://vibat.org. http://www.vibat.org/vibat_uk/pdf/vibatuk_stage1.pdf. Accessed Sept 2015

Wack P (1985) Scenarios: shooting the rapids. Harvard Business Review, Nov-Dec

Wangel J (2011) Change by whom? Four ways of adding actors and governance in backcasting studies. Futures 43:880–889

WEC (2007) Transport technologies and policy scenarios to 2050. Online available at://www.worldenergy.org/documents/transportation_study_final_online.pdf. Accessed Apr 2009

Yearwood Y, Stranieri A (2011) A reasoning community perspective on deliberative democracy. In: Yearwood Y, Stranieri A (eds) Technologies for supporting reasoning communities and collaborative decision making: cooperative approaches, 1st edn. Information Science Reference, New York, pp 94–114

Chapter 4
The Strategic Role of Airfreight, Trade and the Airport: Evidence from Europe

4.1 Introduction

In this chapter, we aim to broaden the understanding of (1) the close relationship between the type of trade of the EU and airfreight demand in recent decades and (2) the growth-inducing effects of airfreight on airport expansion in Europe and in its north-west regions (i.e. the United Kingdom, the Netherlands, Belgium, Luxemburg and parts of Germany and France). The third aim is to scan the patterns in future threats of air cargo firms (i.e. the limits to growth of airfreight volume) through the analysis of the link between airfreight volume, firm strategy and environment (CO_2 emissions pathways).

Airfreight transport has grown on an annual average of 4% in 1980–2017 in Europe's 28 member states (EUMS28) (World Bank 2018); however, the sector faces two key limits: (1) slow airport expansion because of limited land availability within the European Union and (2) high fossil fuel dependency.

Airfreight transport has become a key industry of the current global economy and of the logistics industry. The terms airfreight and air cargo are used interchangeably in the chapter. Airfreight development has been the result of changes of productive chains, of the evolution of the airline industry itself and of changes in the contextual environment under which airfreight operates. The first air cargo shipment of food is recorded in 1928 when 7500 tonnes of fruit, flowers and vegetables were airlifted to London (Kasarda and Lindsay 2011, pp. 211); the first scheduled air mail is recorded in 1911 between London and Windsor, UK (Wikipedia 2013a, b).

The sector's turnover is important for the European economy: it is valued at 131.8 billion euros per year (Eurostat 2012a, b), but a study of the EU logistics market estimates that the logistics operations (excluding in-house operations) can be as high as 878 billion euros (2012) in the EU (Eurostat 2012a, b).

© Springer Nature Switzerland AG 2020
D. Bonilla, *Air Power and Freight*, SpringerBriefs in Energy,
https://doi.org/10.1007/978-3-030-27783-3_4

The EU (27 member nations), as a whole, has an aircraft (cargo only) fleet of 185 (size of <100 k lb (MTOW)),[1] 173 (with a size of >100 k lb), 65 mixed cargo aircraft and 87 aircraft for special purposes. There is also a fleet of 2057 business jets (Eurostat 2012a, b). Passenger aircraft also carries freight. This means the airfreight sector is strategically important for the development of Europe's trade, for its economy, for the geography of airports, and in future, for both the development of cities and where people wish to live (Kasarda and Lindsay 2011).

This chapter fills a gap in the field of airfreight geography and in the transport sustainability field, using evidence for the EU air cargo market as a whole and for North-West Europe. This chapter is organized as follows in Sect. 4.2 we establish the seven key forces of the airfreight sector and describe the literatures on airfreight logistics. Section 4.3 discusses the geographical impact of the changes of airfreight and related logistics activities in North-West Europe in the 1990s to 2000s. Section 4.3 also describes the methodology of the geographical analysis of airports. Section 4.3.2 reviews the interaction between the airfreight sector and its geographical location. Section 4.3.3 introduces a model to estimate the link among firm profits, airfreight volumes, the geography of airfreight and airports. Section 4.4 discusses the historical shifts in the airfreight sector for the European Union, in connection with energy and emissions. In the same section we assess EU airfreight volumes and confirm that the sector impacts are growing (measured by its energy use and its emissions).

Section 4.5 discusses the future of airfreight sector in the light of contextual and contractual environments surrounding the airfreight sector as well as the changing economic structure of airfreight and logistics.

4.2   Airfreight and Logistics: Key Forces

4.2.1   The Seven Key Forces

Logistical and airfreight transport activities have been supported by seven factors: first is the globalisation of Europe's economy since 1970. The second factor is trade, both intra and extra types, which is closely linked to export and import activities of the EU economy. The third is the adoption of the hub and spoke network by airlines. The fourth factor is the deregulation of national economies. The fifth is the adoption by private firms of 'Just in Time' practices. The sixth is the spread of electronic commerce (e-commerce) and broadly speaking technological change. The seventh one is the spatial dispersion of production lines and distribution. We explain these forces in turn below. Figure 4.1 depicts the time trend of airfreight growth.

---

[1] MTOW: Maximum take off weight.

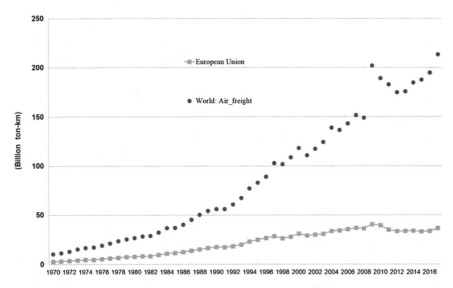

Fig. 4.1 The time path of airfreight: World & European Union (billion t-km). (Source: World Bank (2018))

As Fig. 4.1 shows there have been a few episodes of declining airfreight activity on a global scale. The declines were recorded only in 1983, 1991, 2001, 2004 and 2009–2012. The decline in 2009 was almost as large as that of 2001. These declines in airfreight volume were unparalleled in recent decades on a global and EU scale. In 1980–2017, EU airfreight has grown by 4% annually on average (World Bank 2018) just below the rate of growth of global airfreight of 5.7% annual average using t-km (World Bank 2018). Figure 4.1 depicts these changes in airfreight. Airfreight volume of the EU represents about 1/6 of the world's total in 2017; the share of the EU in the global airfreight flows continues to fall. The growth in EU airfreight is slowing down.

The second driver that explains the growth of airfreight is trade. Trade activity in Europe was influenced by the expansion of the European Union which lowered import tariffs of goods and services and encouraged greater European integration. The export and import (External trade) share by transport mode is given in Table 4.1. The third factor is the adoption of hub and spoke networks after the liberalisation of intra EU market. A hub network is defined 'as a system of connections arranged like a chariot wheel in which all the traffic moves along the spokes connected to the hub at the center' (Wikipedia 2013a). Memphis International Airport, in the United States, is the quintessential modern airfreight hub. The adoption of hub and spoke networks, within the airfreight sector, has increased both freight volumes and connectivity; many airfreight/network carriers serve the local market and are connected to other airports at the same time. Due to the consolidation of the EU and US air transport industry, more and more network carriers operate out of multiple connect-

Table 4.1 The role of airfreight in external trade: EU 28 MS (2016) (Values in %)

	Exports value (%)		Import value (%)		Export quantity (%)		Import quantity (%)	
	2002	2016	2002	2016	2002	2016	2002	2016
Air	28.5	28.9	22.4	24.5	1.1	2.3	0.2	0.3
Ocean	43.3	47.6	43.4	50.8	72.8	80.8	67.4	73.4
Road	17.6	18.1	15.1	14.9	14.5	12.5	3.2	4.1
Rail	1.7	1.2	1.3	1.2	4.5	2.6	4.8	4.3
Others	8.8	4.2	17.7	8.5	7.1	1.7	24.4	18

Source: Eurostat (2019). Others: inland waterways, pipeline and self propulsion. Excludes intra EU trade

ing hubs (Burghouwt 2013). The hub and spoke network has driven the expansion of the airfreight sector and not only that of the passenger sector.

The fourth driver is the current deregulation of the economy which responds to the crisis of the post-war Fordist system (Dicken 2003).[2] The fifth relates to the introduction of new logistics practices, for example, Just-in-Time (JIT) schedules which reduced warehousing costs together with technology advances in modern transport.

The sixth driver is the invention, and adoption of the steam engine, followed by the ICE engine, the jet turbine and more recently by the widespread adoption of modern information technology (IT); this factor operated in tandem with the need for time critical products. The seventh factor is the spatial dispersion of production lines and distribution which is induced by IT and a rapid increase in international trade (Carroué 2002; Dicken 2003). The adoption of IT has spurred the growth of e-commerce, and express mail in warehouses, in trucks and in entire supply chains has also been a driver of the airfreight market (Kasarda and Lindsay 2012).

In this context, the airfreight mode has experienced fast growth. Its activities were first limited to the exchanges of essential goods or for high value-added goods. Airfreight is now part of many supply chains and of regular exchanges services (Hesse and Rodrigue 2004). This evolution has two origins. First, the supply chains have become more complex and rapid, which has encouraged a more frequent use of rapid exchange under the practice of Just-in-Time (JIT) methods. Second, the price of airfreight has declined both in absolute and in relative terms, in fact real air freight rates have fallen by 3% annually which means more goods can be freighted via air (Leinbach & Bowen 2004).

---

[2] The fordist system is the economic model that prevailed in Western countries during the post WW II era until the 70's. It was based on the Welfare State, a regulated economy and on a the constant growth of wages, financed by a) constant gains in productivity, b) reducing production costs and c) by the expansion of both of mass consumption and of the middle class.

4.2.2 Intermodal Competition and External Trade of the EU

In this section we discuss the modal share (air, ocean, road, rail, inland waterways or IWW, pipeline and self-propulsion) of external trade for the EU 27 member states.

Europe's international trade is part of this evolution (see Table 4.1). In quantity terms, air transport represents a small share of EU imports (0.3%) and its exports (2.3%) in 2016. However, this share is far higher in terms of value: 24.5% of imports and 29% of exports in the same year. The ocean mode represents the largest volume of external trade in physical units, but this contribution falls in value units for both imports and exports. The air mode share is higher than the share of the road mode using the value of exports (Table 4.1).

Less important modes show a different picture if we examine trends in Eurostat (2012a, 2012b). The share of freight moved via pipeline is higher than that of the IWW mode; the latter mode is relatively unimportant in terms of both physical freight movement and its value (not shown in Table 4.1). Pipelines move little freight in the export balance but do move freight in the import balance account. The IWW and the pipeline modes play apparently little role in the total freight market, but the sector ought to save significant amounts of fuel since the modes reduce the number of trips by air, truck and train. The pipeline mode is key for importing energy products such as natural gas but less so for exporting other commodities.

The penetration of the airfreight mode by airport at NTUS 2 level is tabulated in Table 4.2 (Eurostat 2015) in physical units for the various European regions. Airfreight services are increasingly used for moving goods over long distances when there is demand for high value and for perishable goods (Eurostat 2015).

In general a few regions in the EU dominate airfreight transport. Four regions: Darmstadt (Frankfurt), Noord–Holland, Ile de France and the outer London regions concentrate most of the airfreight movements inside the EU region: their combined share is 59% of total airfreight moved within the EU. Frankfurt airport is the leading European airport for freight with a market share of 16% while the three other regions hold shares above 10%. The distribution of airports reflects the geographical distribution of the highly developed regions of Europe as far as high technology and manufacturing goods production.

Both imports and exports by airfreight are largely dominated by manufactured goods (Eurostat 2012a, b) i.e. textiles and mechanical parts, machinery and appliances as well as engines have a combined share of 83% in import bill and a 75% one in export one in 2012 (Eurostat 2012a, b). On the import side, food products represent only 15% of total import bill by weight value (tonnes) and 1% of the export one. The changes, as described above, reflect how Europe has integrated into the global economy.

The over-representation of the aircraft mode for exports (Table 4.2) reflects a European industrial economy that has turned to the production of goods with higher value added (Dicken 2003). The faster growth of exports illustrates the same phenomenon: the specialization of European industry towards high value-added goods.

Table 4.2 Airfreight transport and top 20 European Union Regions by NUTS level 2 region

	Volume of airfreight and mail loaded and unloaded (thousand tonnes)	Volume of airfreight and mail loaded and unloaded airfreight volume (tonnes per 1000 inhabitants)	Volume of airfreight and mail loaded and unloaded (tonnes per km^2)
EU-28	13,384	26.4	3.0
Darmstadt (DE71)	2095	548.1	281.4
Noor-Holland (NL32)	1566	571.2	382.8
Ile de France	1559	129.9	129.8
Outer London	1514	295.8	1195.1
Leipzig	877	889.0	221.2
Koln	722	166.6	98
Luxembourg	673	1224.3	260
Lombardia	566	56.8	23.7
Prov. Liege	534	486.1	138.3
Prov. Vlaams-Brabant	379	341.7	180
Comunidad de Madrid	367	57.1	45.7
Leicestershire, Rutland and Northamptonshire	297	170.2	60.4
Oberbayern	288	64.4	16.4
Essex	236	134.2	59.8
Niederosterreich	190	116.8	9.9
Helsinki-Uusimaa	187	117.9	19.5
Lazio	158	26.9	9.2
Hovedstaden	137	78.3	53.7
Koblenz	133	90.2	16.5
Southern and Eastern	127	37.7	3.5

Source: Eurostat (2015)

Finally, the fast growth of imported fresh products illustrates the globalization of food supply chains of supermarkets. Relying on exports of high value-added goods allows the EU to compete in the global market. This evolution has led to profound changes in the structure of airfreight and logistics (Bowen and Leinbach 2004; Hwang and Shiao 2011).

The previously cited regions (Outer London, Darmstadt, Noord Holland, Ile de France) also stand out as the top airfreight intensive regions (airfreight volume divided by territory); however, Leipzig, Luxembourg and Liege airports are also key centres. The same regions also stand out of the list as far as freight if this is measured on a per capita basis. Future developments in the high-tech industry of these regions are likely to induce growth on airfreight activity in Europe.

4.2.3   Literature Reviews of Airfreight

The literature on the airfreight transport sector in relation to new logistics practices and to its environmental impacts are scarce. As Alkaabi (2011) maintain, the literature has paid more attention to passenger traffic than to airfreight traffic. This is compounded by the fact that the statistics on airfreight, logistics and their environmental impacts are quite scarce too.

Several authors study the factors that explain the growth of airfreight. Alkaabi (2011), Button and Yuan (2012), Sun et al. (2010) and Kasarda and Green (2005) highlight the link between airfreight and GDP. These authors demonstrate that cargo volumes rise above the rate of economic growth in general because of the internationalisation and the acceleration of freight exchanges and the need for exchange of high value products. Thus, they estimate that this growth will continue in the next decades, at an even faster pace than that of passenger services (Kasarda and Green 2005; Kiso and Deljanin 2009). Oum et al. (1999) examine the regulatory events that surround the growth of airfreight; They highlight the Chicago convention and the creation of the EU single aviation market which aims to increase airline competition and to lower airfreight rates. In this context, it has been argued that the liberalization of airfreight exchanges led to a concentration of this activity in the hands of a small number of big international firms (Zhang and Zhang 2002; Yamaguchi 2008). These firms organize their traffic around continental hubs. The preferred structure of these firms is the single hub organisation while multi-hub structures are used for the US market (Reynolds-Feighan 1994; Zhang and Zhang 2002; Oum et al. 1999).

The link between airfreight and logistics is harder to establish empirically. Alkaabi (2011), Oum et al. (1999) and Kiso and Deljanin (2009) argue that airfreight may be a driver for the location of industrial and logistics activities. However, those authors highlight the need to differentiate integrators' hubs and those serving urban airports. Logistics activities around integrators hubs are few, because they are only related to the transhipment activities of the integrator. Negrey et al. (2011) illustrate this with the case of Louisville, the North American hub of UPS.

On the one hand, urban airports generate more logistics services because urban markets are the main destination and origin of airfreight. Goods must be handled, warehoused and expedited in these places. Experts advocate the creation of aerotropolis for these places, i.e. airports that integrate high value services, logistics capabilities and industrial activities related to airfreight (Alkaabi 2011). On the other hand, the presence of freight forwarders and other logistics firms is key for the activities of classic airfreight companies, since they do not assume the final deliveries and logistics services to their clients (Gardiner et al. 2005; Sun et al. 2010). This distinction remains a theoretical one, however, for two reasons: (1) the absence of empirical evidence of this problem and (2) the difficulty in differentiating between transit traffic and the one that is originated from or destined for the local market

(Alkaabi 2011). In the same way, Hesse and Rodrigue (2004) point out the importance of airfreight for the new logistics but without providing any empirical evidence. Kasarda and Lindsay (2011) provide historical evidence on the impact of airfreight on logistics, and on the geography, of the aerospace industry in California.

As far as the airfreight environment nexus, the sector has been studied (McKinnon et al. 2015) but without reference to trade flows. Despite the fact that this transport mode is a highly polluting one in terms of CO_2 emission per tonne kilometre, the impacts of new logistics organisation (using hub-spoke airports) on the environment have not been widely analysed in the literature. The literature shows overwhelmingly the negative environmental effects of hub and spoke exceed the positive ones (Pels, 2008). The subject is difficult to research because data on airfreight are limited in availability: aircraft in use, volume of freight per destination or the impact of the combined flights (passenger and freight). However, Kohn and Brodin (2008) shed light on the negative impact of hub and spoke network organization of air freight flows in terms of CO_2 emissions produced by short term flights.

4.2.4   The Structure of the Airfreight Industry in the EU

Recent studies on the structure of the airfreight industry have identified various players. Three types of players share the air cargo market (Alkaabi et al. 2011; Zhang and Zhang 2002). The first type of player is the freight integrator, i.e. the global companies specializing in parcel express transport (DHL, TNT, UPS, FedEx). These companies have responded to the demands of the economy by integrating new industrial logistics services, such as storage, land transport, shipping and supply chain management (Kiso and Deljanin 2009). They are the winners of the deregulation of airfreight exchanges because they can answer to the requirements of the JIT economy (Zhang and Zhang 2002).

The second type of actors is represented by the many smaller airlines specialized in freight transport (Kiso and Deljanin 2009). Their scope of activities is small. Finally, the 'classic' airlines are those primarily focused on passenger transport, but which also provide air cargo service. The organisation of these airlines remains dictated by their passenger activities, so these companies depend on freight expeditors and other logistics actors for the delivery to their final clients (Reynolds-Feighan 1994; Kiso and Deljanin 2009). The organisation of the airfreight industry is dual (Zhang and Zhang 2002): Integrators prefer hub and spoke schemes, because of the goods they move: parcels and small packages that are easily handled. Other airfreight companies focus on direct flights, because it is more reliable and quicker, especially for fragile, perishable or unconventional goods. So airfreight alliances remain rare because of the lack of transhipment and hubbing activities.

4.3   The Integration of Airfreight in Logistics Organisation: A Geographical Analysis

In this section, our aim is to identify the geographical implications of the changes of the spatial organization and the role of air transport in today's economy for actors of air cargo and of logistics. Our analysis is based on an empirical study of the localization of logistics and airport infrastructure across North-West Europe. The section is organized into three subsections: (a) data and methodology, (b) results and (c) discussion.

4.3.1   Materials and Methods

This section is based on Eurostat data and on a study in Strale (2010). The latter used census data is used of industrial sites for 500 logistics companies, representing 4000 sites in North-West Europe (see Fig. 4.2). Strale classified the data according

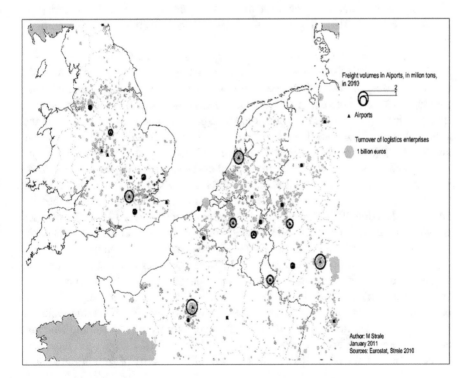

Fig. 4.2 Airports and logistics companies' locations in North-West European countries. (Source: Strahle 2010. Unpublished study)

to their business activity: International transport, road transport, storage, supply chain management and fresh food logistics, among others.

The traffic statistics include airports that have moved more than 100,000 tonnes in 2010; the traffic data is sourced from Eurostat.[3] Data on these airports are ordered following three categories by (a) air cargo hub of a global integrator (i.e. DHL, TNT, FedEx or UPS), (b) non-airfreight hub and (c) major airfreight company (i.e. Lufthansa Cargo, Cargolux or Air France).

4.3.2   Results

Our results are organized in two steps. First, we study the geographical changes at the airport level, by studying their hierarchy, the type of air cargo players that are present and the existence of concentrations of logisticians in their neighbourhood. Second, we study how airport location affects air cargo and how logistics companies feed on each other and contribute to airport expansion.

Figure 4.2 illustrates the geographical locations that are examined to explain the geography of airfreight and logistics activities. Figure 4.2 compares the turnover by industrial establishment of logistics firms at the municipality level. The data is based on Strale (2010). It gathers the biggest logistics companies operating in Europe.

The census data obtained by Strale (2010) includes around 500 companies and 5000 logistics establishments. Logistics establishments are classified into eight types, according to firm activity: general logistics and warehousing, general industrial logistics, automotive logistics, cold logistics and supermarket logistics, general transport logistics, express transport, road transport and international transport, and supply chain management. This classification is based on Samii (2001) and Carbone and Stone (2005). The data on the turnover of these logistics activities is also known.

Cargo airports and their traffic are located at the municipality level. North-West Europe is the main focus, and it is based on the ESPON and Eurostat definition of this space. In North-West Europe, the main cargo airports are located near the biggest cities, London, Paris, Amsterdam, Brussels or Frankfurt (see Fig. 4.2). Logistics activities are concentrated around the main urban areas and near the biggest North-West European ports and airports. Cargo airports have grown steadily since 1990 (see Table 4.3). Their traffic is generated by the three types of air cargo actors i.e. airfreight integrators, full cargo companies and passenger airlines (Bocquet 2009). However, over the last decade, the four airports, cited above, are no longer the leaders in terms of capacity expansion. A few secondary airports (Cologne-Bonn or Liege) have grown faster. The hub of freight integrators, Paris, Liege and Cologne are now leading the growth airport capacity in North-West Europe; the latter airports take advantage of the tremendous growth of the activity of freight integrators

---

[3] Overview of the freight and mail air transport by country and airports database, Eurostat website, *epp.eurostat.ec.europa.eu,* statistics database section.

Table 4.3 Major North-West European airports and their activity (thousands of tonnes)

	1990	1995	2000	2005	2011	% growth (tonnes) 1995–2011	% growth 2000–2011
Amsterdam Schiphol	604		1222	1449	1523	152%	25%
London Heathrow	695	1031	1307	1305	1484	44%	14%
Koln Bonn		312	423	636	726	132%	72%
Brussels		464	687	702	475	2%	−31%
Manchester		90	116	147	115	128%	−1%
Frankfurt	1176	1521	1589	1892	2133	58%	34%
Paris Charles de Gaulle			1517	1767	2088		38%
Liege	0	0	270	325	674		150%
Maastricht	0	0	0	75	98		
East Midlands				292	304		
London Gatwick			318	222	104		−67%
London Stansted			260	254	204		−22%
Frankfurt Hahn	0	0	25	107	286		1044%
Luxembourg		286	499	742	657	130%	32%

Source: Eurostat (2012a, b), Wikipedia (2013b) for air cargo routes

Table 4.4 Major North-West European airports and their activity (thousands of tonnes)

Airport	No. of cargo routes	Population Within 100 km (million people)
Amsterdam Schiphol	91	9
London Heathrow	80	33
Koln Bonn	70	16
Brussels	51	8
Manchester	19	11
Frankfurt	84	35
Paris Charles de Gaulle	99	22
Liege	88	5
Maastricht	15	5
East Midlands	42	10
London Gatwick	19	33
London Stansted	37	33
Frankfurt Hahn	17	8
Luxembourg	76	3

The catchment area is shown in population size; connectivity increases airfreight. Source: Eurostat (Eurostat 2012a, b), Wikipedia (2013b) for air cargo routes

and of the establishment of a hub and spoke airline organization that generates a large volume of trans-shipment.

Secondary airports that are not hubs, i.e. Gatwick, Stansted, Maastricht, are following different growth strategies, but they remain limited to low traffic, generated

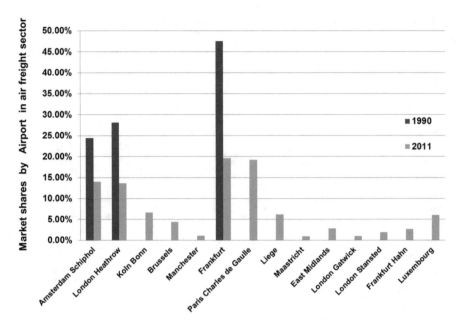

Fig. 4.3 Market shares for selected years (volume of airfreight) of airports in 1990 and in 2011. (Source: refer to Table 4.3)

by the small airfreight companies. Table 4.3 shows the historical changes in freight activity by regional airports in Europe, and Table 4.4 shows the correlation between connectivity, population density and airports. The correlation between population centres and airport connectivity is low (10%) contrary to theory of economic geography. However using data across time periods should show a stronger positive correlation.

Figure 4.3 shows the market shares (in tonnes) of the same airports (listed in Table 4.3) for 2 years; the impact of competition from smaller airports can be seen in Fig. 4.3. Paris Charles de Gaulle clearly has amassed market share in the sector; the latter is one of the top cargo airports in the world, but it is behind Asian airports such as Hong Kong and Singapore's Changi.

In 2018, the busiest top ten air cargo airports in the world are Hong Kong, Shanghai's Pudong, with three in the USA (Memphis, Ted Stevens in Alaska and Louisville); and only two are in Europe (Frankfurt and Charles de Gaulle). Dubai is also in the top-ten list (Wikipedia 2013b). In future, Beijing airport will handle five million tonnes of cargo a year staying ahead of any other airport in the world. The competitive pressures on European-Asia routes are, therefore, increasing for European air cargo airports.

Within North-West Europe, London Heathrow, Frankfurt and Amsterdam Schiphol airports are losing market share, and Frankfurt no longer has the commanding lead out of all airports. This points that the market concentration of the leading cargo airports in Europe is declining. One reason for this is that airfreight is

Fig. 4.4 Geographical concentration of logistical activities in North-West Europe. (Source: Strale 2010)

helping spread the arteries of supply chains (and freight firms) beyond the leading airports (Fig. 4.3).

By using the method of the nearest neighbours first used by Pinder and Witherick, in 1972, Strale identifies the high density of logistics activities (see Fig. 4.4). Strale finds three levels of concentrations of logistics activities. Concentrations of logistics activities are located around the biggest North-West European cities and urban areas: Paris, London, the Ruhr area, the Midlands and the Benelux. Port cities also concentrate an important logistics activity. A small number of concentrations of logistics activities surround the North-West European airports; they are coloured in grey (Fig. 4.4).

By matching geographical concentration of logistical activity with the volume of freight handled by airports, it is possible to differentiate these airports. To do this, experts estimate a nonlinear regression for airfreight traffic levels; the latter is

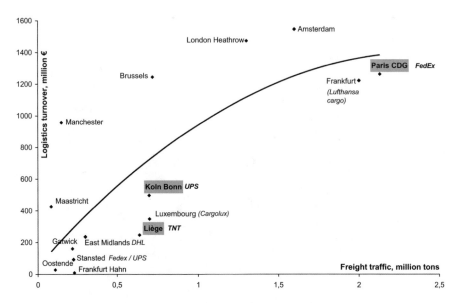

Fig. 4.5 Regression between logistic activity and volume of airfreight of airports in 2010. Liège *TNT:* principal hub of TNT; East Midlands *DHL:* secondary hub of DHL. Source: Bocquet (2009) using data in Eurostat (Eurostat 2012a, b)

explained by the turnover value (millions of euros) of logistics activities (see Fig. 4.5).

The following non-linear regression model (Strale 2010) can be used to explain the behaviour of airfreight in relation to turnover.

$$\text{Airfreight}_i = \alpha + \beta_1 \log (\text{turn})_i + \beta \log (\text{turn})_i^2 + \in_i \qquad (4.1)$$

where *i* stands for the number of airports; *airfreight* is the volume of airfreight traffic per year at each airport and it is a proxy of airport services, *Log turn* logistics turnover. Epsilon stands for non-observed effects on airfreight. The results of the regression are reported in Table 4.5. The quadratic term of variable *turn* captures how much airfreight increases for every additional euro of turnover at each airport. On average, a unit increase in logistics turnover will lead to an increase in airfreight (proxy for airport services).

Strale (2010) shows that two variables are positively correlated: the logistics activity of firms grows following an increase in airfreight traffic volumes. However, the evidence furnished by Strale (2010) and Bocquet (2009) shows this relationship tends to weaken for the biggest airports: it may be an indication of saturation of logistics activities around these airports or of spatial limits surrounding the airports.

Regarding Eq. (4.1) above two comments are in order. First we do not estimate the above but use the model structure to explain what variables are isolated from the overall transport economic context. Second, the model (1) works as a guide for modelling freight transport and logistics turnover activity at a given location.

Table 4.5 Type of airport and logistics activity at the airports

	Airports			Outside airports	
Activity	Urban	Hub of integrators	Small/full freight	Urban centres	Suburbs
Freight integrator	+	+++	+	++	+
Regular airlines	+++	−	− − −	+	− −
Small airfreight company	++	++	++	+	− −
Logistics SCM	+++	+	+	++	+
Other logistics	−	−	−	−	+++

− − −: Highly unpreferred, − −: unpreferred, −: slightly unfavourable
+++: highly preferable, ++: preferable, +: slightly favourable
Sources: Strale (2010 analysis)

In theory if turnover induces airfreight activity, this would confirm other studies (Bel and Fageda 2005): Increases in airfreight traffic (services) should encourage logistics turnover (activity). Our results reflect the influence of airfreight activity on urban logistical activities.

A trade-off appears: Airports housing a hub of one of the leading express mail companies have a smaller volume of logistics than other airports with a comparable airfreight activity. Indeed, hubs create significant traffic, largely composed of goods transfer via truck (Hall 1989). However, these hubs hardly induce activity except for the global express mail operator itself (CIRIEC 2002). Luxembourg and Frankfurt are hubs for Lufthansa Cargo and Cargolux, and both are part of a quite similar pattern (Bocquet 2009).

Airports of the largest cities represent important markets for air transport. Many customers are found in these metropolitan areas. Goods that are handled in these airports are generally destined for, or originate from, the local market, inducing logistics activity (Noviello and Cromley 1996). These factors explain the higher logistics volume in the vicinity of these sites (Bowen and Leinbach 2004). In contrast to airports serving the large city, smaller airports are located further away from urban centres, such as Oostende, Gatwick or Frankfurt Hahn, and they hardly induce any logistics activity. These terminals are used by some companies specializing in air cargo, taking advantage of low costs, and do not creat hubs (Gardener and Ison, 2008).

Given the above discussion of airport location, a pattern appears involving three types of terminals. The first one is the city airport, which is well established, and is a major logistical node. City airports have an average growth of traffic, based on both firms of freight integrators and of firms of classical airlines. The second type is the hub integrator that shows high growth and is highly specialised but that induces hardly any logistical activity. The third type is the secondary airport without any hubbing activity, which has a marginal role in terms of air cargo and of the logistics business.

The actors that cluster around the main cargo airports are active in (a) international transport (freight and handling) or supply chain management and (b) in the express transport field. These actors are regularly using air transport on behalf of their clients or for their own business. In contrast to these actors, logistics companies, that are specialized in ground transportation or storage, are not attracted by the airports; examples include the automotive and the logistics operations of supermarkets. As a result of the economic activity, the land is more expensive in these places and the available space is reduced inhibiting firms to rely on airfreight services (Hesse and Rodrigue 2004).

Meanwhile, players i.e. express transport operators, in the air cargo sector have been active in expanding their logistics activity. Following the DHL model, all of these players have integrated new logistics services to broaden their range of activities and become global logistics providers (Bocquet 2009). To achieve this end, the preferred strategy was to associate with other companies like former public postal groups or important freight transport actors. This strategy has led them to the top of the European logistics ranking. There was a clear strategy to expand their activity outside the airports for both economic and geographical reasons. The largest conventional air cargo companies such as Cargolux and Lufthansa Cargo have followed similar strategies, by expanding services to land logistics (Noviello and Cromley 1996).

Other cargo airlines have not expanded their logistics activities as much as the above described transport operators. This is justified by several factors. First, as far as passenger airlines, freight is often a secondary activity, which does not justify an expansion strategy. Second, smaller air cargo companies generally occupy specific niche markets that do not justify massive investment (Gardiner and Ison 2008). However, these companies do not stay completely inactive. For example, these companies have built alliances with other cargo airline firms, in order to share equipment and expand the destinations and the services for their customers. A third factor is the increasing separation between freight and passengers. Freight administrative departments of passenger companies have been progressively outsourced and merged together to form larger independent groups.

4.3.3   Implications for Types of Airport

According to the established literature, three types of airports can be differentiated: city airports, airports serving as hubs for integrators and as secondary airports. This is shown in Table 4.5.

Although the data of Table 4.5 seem outdated; it does lay the basis for future research among logistics, airfreight expansion and airport type. City airports concentrate the activities of passengers and companies and are the geographical point of departure, or of arrival, of many express mail integrators services. The fact that city airports serve metropolitan markets makes them particularly attractive for logistics companies. Therefore, city airports are major logistical nodes across North-West Europe.

The hub airports are located in remote peripheries of urban centres or in the interface among major metropolitan areas. These are the airports whose recent growth has been fastest. These airports benefited from the growth and the expansion of services of global freight integrators. However, this high traffic does not result in the development of induced logistics activity. Indeed, these are mostly transhipment flows, between aircraft or between airfreight and other transport modes, which only pass through these platforms. Finally, the secondary airports are in intermediate situations. Being more remote of cities, secondary airports adopt niche strategies. These airports are especially used by low- or medium-scale cargo airlines and do not constitute logistics nodes.

Meanwhile, world level players of airfreight have adapted in various ways. For instance, freight integrators have undergone the most profound changes: (a) they now cover the full range of logistics services and (b) have broader coverage of the territory, whether in the metropolitan areas or in urban peripheries. Traditional airlines are generally not focused on freight and have adapted more slowly. This occurred via rationalization of their supply chains through alliances or extensions of services to high value-added logistics. In this context, traditional airlines focus their activities in the metropolitan centres. Finally, smaller airlines have changed little and often experience stagnation or decline.

The logistics companies that are mostly attracted by airports are the ones that operate in the field of supply chain management. These firms need geographical proximity to markets, and so they tend to be based in metropolitan locations. In this context, airports are attractive because they form international exchanges nodes. The other logistics firms, with operations for freight transport or for storage, are less attracted to airports. The need for cheap land and for access to roads means logistics firms prefer to be located within suburbs. These logistical activities have contributed to the growth of airfreight in the EU as a whole and have accelerated directly or indirectly the growth of energy use and CO_2 emissions growth of airfreight.

4.4   Impacts of Airfreight on the Environment

We describe the impacts of airfreight movement on energy, fuel burn level of aircraft, the environment and other impacts of airfreight activity. Some solutions are suggested to minimize CO_2 emissions generated by airfreight activity.

4.4.1   Impacts of Logistics, Its Evolution and the Environment

Given the above-described growth of airports and associated freight volumes (Fig. 4.1) and changes in strategies of airfreight firms, we now examine its environmental impacts. There are no official statistics on the local pollution and carbon emissions of airport logistic activities; Sect. 4.4.3 shows our CO_2 emissions estimates of airfreight movements of the EU 27 MS. The following section discusses the environmental impact qualitatively and quantitatively.

World aircraft fuel consumption (all countries considered) has grown above other freight modes in the last 10 years. Two broad parameters explain energy use and emissions of aircraft fleets in North-West Europe: (1) physical determinants of aircraft operation and (2) consumer demand for airfreight services increasing aircraft payloads. Larger payloads (freight volumes) of the aircraft respond to higher volumes of intra EU and extra EU export, and import, expansion (Table 4.1). Energy use can also be explained by the absence of available fuel substitutes for aircraft engines. Further technological issues, determining the amount of energy use, include specific fuel consumption, the lift to drag ratio and structural weight. The Breguet equation can be applied to estimate aircraft (airfreight) fuel use using data on fuel burn rates (Lee et al. 2011).

4.4.2 Energy Use

Aircraft fuel consumption in the European Union is about 50.4 million tonnes of oil equivalent (MTOE) (Eurostat 2012a, b), and most of the growth of consumption comes from the top 10 European countries. Jet fuel, or kerosene, powers most of the world's aircraft fleets as well as those of North-West Europe, and the sector is financially vulnerable to jet fuel price changes: A doubling of fuel price leads to an

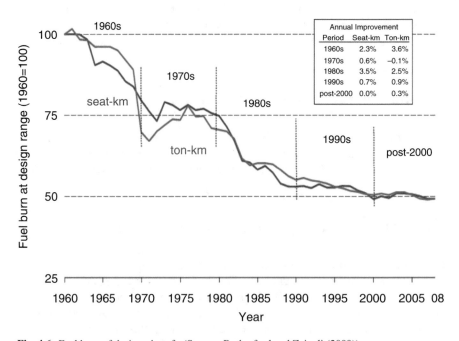

Fig. 4.6 Fuel burn of design aircraft. (Source: Rutherford and Zeinali (2009))

11–26% rise in aviation costs (Hummels 2009). As of today, in aviation there are no substitutes for jet fuel.

World aircraft fuel consumption, governed partly by fuel burn rates, is 264 MTOE in 2006 (IEA 2012), and it is projected to grow by 6% per year once the global economy recovers (Boeing 2009). At this moment, there are no ways to mitigate this level neither of fuel consumption nor of CO_2 emissions of aircraft fleets. Historically, energy intensity of airfreight activity has fallen by about 60% since 1970 (IEA 2012) partly as a result of fuel burn gains and of temporary high jet fuel prices. Figure 4.6 plots the fuel burn since 1960s. In that decade positive improvements of fuel burn rates were achieved but little progress has been seen in recent decades.

Fuel burn rates did improve from the 1960s and 1970s and 1980s contributing to lower fuel consumption. Fuel burn rates, however, have not fallen as much as needed, especially in the 1990s and 2000s decades. Major improvements in fuel burn did appear in the decade of the 1970s (Fig. 4.6). The slow improvements, the larger aircraft fleets (North-West Europe included) and the larger exports indicate that the aircraft industry, and the airfreight sectors, will continue to increase fuel use and be highly dependent on fossil fuel resources.

We describe what could happen in the air transport sector, without providing detailed prescriptions on how to remodel the airfreight sector. The airfreight sector will have to introduce energy efficiency measures (more on Sect. 4.4.3) in order to achieve carbon emissions cuts; but the latter may be less effective if allowance is made of the energy demand that can rebound. A recovery in energy demand of aircraft or a 'rebound' results from improved burn fuel efficiency of aircraft in operation or other technological measures such new plane designs, lighter materials or biofuels (see IATA 2013). This rebound effect can be as high as 12% of aircraft fuel consumption after a reduction in jet fuel prices (Barker et al. 2009). This means that a 10% improvement in fuel burn rates will not decrease fuel consumption by the same amount but by 8.8% [10 * 0.12 = 1.2] and [10% − 1.2% = 8.8%]. If the reduction in energy use is equal to energy efficiency gain, the rebound is zero but this is not the case.

The activity elasticity for aircraft can lead to a 49% change translating into an increase in aircraft fuel consumption after a recovery in the world economy (Barker et al. 2009). The possibilities to substitute fossil fuel in this sector are rather limited and so are the fuel burn efficiency gains.

4.4.3 Emissions

The high volume of economic activity (Fig. 4.1, Sect. 4.2), with the associated export and import demand for goods, and with the increased the dynamism of logistics firms in the entire European region (North-West Europe) inevitably widen the environmental impacts and worsen air quality.

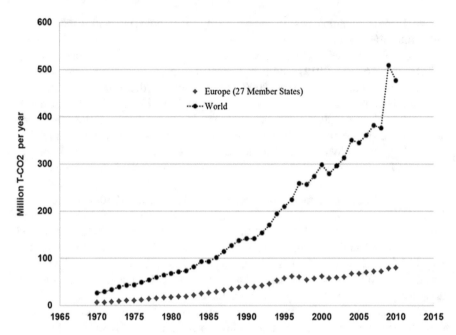

Fig. 4.7 Carbon dioxide (CO_2) emissions of airfreight. CO_2 emissions are estimated using IPCC (1999) emissions factors for airfreight movements and data of airfreight in World Bank (2018). The carbon intensity is 700 G-carbon emitted per tonne-km

This also increases local emissions and CO_2 emissions, as Fig. 4.7 depicts. This tracks the historical behaviour of CO_2 emissions in 1970–2010. Figure 4.7 depicts the time trend of emissions of CO_2 per year for airfreight movements of EU 27 MS. In 1970–2010 years, the figure shows that the emission have doubled within the last two decades. The recession of 2008–2010 does not show a decline in emissions of CO_2 unlike the global emissions level, which does show such decline.

Global CO_2 emissions of air travel stand at 12% of global transport emissions (IPCC 2012). In addition to aircraft operations, the activity of logistics firms located in the surroundings of airports will tend to increase indirectly CO_2 and local non-GHG emissions (such as nitrogen oxides, or NOx, ozone, methane, water, contrails and particles) through the use of a variety of machinery.

A challenge facing the airfreight sector, should it intend to cut its emissions during aircraft operation (and during airport time), is the unit CO_2 intensity of aircraft (this point is discussed above) which tends to be higher in the medium haul trips than in long distance flights (IPCC 2012). Unit emissions of aircraft are much higher than any other mode especially at the take off stage (Gilbert and Perl 2008). A further challenge is to determine how to price (or to tax) the CO_2 emissions of airfreight by introducing an emissions trading system alongside the EU emission trading system (ETS) emissions (this is described below).

Compounding the environmental effects of air cargo is the radiative forcing (this measures the warming effect of emissions) of CO_2 emissions, at high altitude, which is three times that of CO_2 emitted on the ground level (RCEP 2002).[4] Some even claim this level of forcing is even as high as five times. The warming effects of some aircraft emissions are highly uncertain. In other words, current logistics practices, through the demand for airfreight services, produce a variety of environmental emissions and alter the atmospheric chemistry, and this means solutions will need to be found that will need to go beyond technological improvements. This requires some form of airfreight demand reduction in the long run which can only mean that the logistics sector, and airfreight, will need to operate in a global economy within carbon emission limits.

There are only a few policy initiatives to reduce CO_2 emissions of airfreight. IATA (2013) argues 'of the four pillars, technology has the best prospects for reducing aviation emissions'. But the IATA (2013) does not make any reference to the possibility of rebound effects of airfreight demand, nor in airfreight energy use.

Besides technology, an economic tool to curb emissions is the inclusion of the sector in the European ETS (EC 2012). The sector is to be included in the ETS starting from 2012. The ETS will allow CO_2 emissions trading in the European stock markets. This policy, however, needs the accurate allocation of emission cuts among nations, a guaranteed CO_2 price to be effective and liquid markets. The allocation of emissions reductions among those firms responsible for emissions reductions can be complex because of the broad mix of economic agents i.e. airports, airlines, aircraft owners, logistics firms, aircraft makers and fuel refiners. Including airfreight in the ETS will impose a price on emissions. It is unknown how the ETS will reduce both airfreight volumes and, in the long run, airport expansion. Other policies to decrease emissions include new technologies (biofuels above), operations and infrastructure.

Although additional work is needed to understand trends of jet fuel consumption of airfreight activity, this analysis helps to understand that it is difficult to control the growth of jet fuel consumption so long as (1) fuel burn rates do not improve significantly, (2) overall volumes of cargo handled by air cargo firms and logistics companies continue to expand and (3) so long as there is an increase in world trade. Policy initiatives such as pricing CO_2 emissions of aircraft may not be enough to offset the increase in CO_2 emissions from airfreight activity.

4.5   Discussions: The Future of Airfreight

Given the need to predict the future volume of airfreight and its impact on CO_2 emissions, a brief analysis of the future of airfreight is called for. In this section, a conceptual analysis of airfreight is described using systems theory (Ackoff 1974; Ramirez 2010) to imagine possible futures of airfreight. Traditional approaches for

---

[4] Radiative forcing is the change in the energy balance of the earth atmosphere system in watts per square meter (W m^{-2}); a positive forcing implies a net warming of the earth and a negative value implies cooling (Babikian et al. 2002).

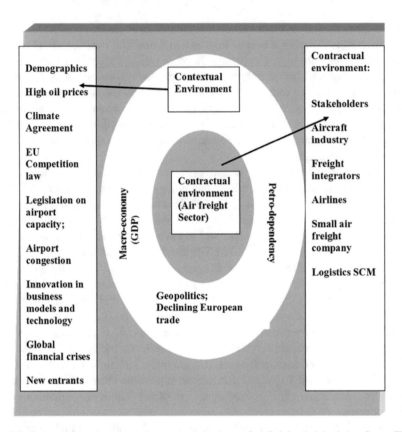

Fig. 4.8 Contextual environment shapes the behaviour of airfreight and logistics firms. This shapes future evolution of the sector. The contractual environment is under the influence of the outside forces (context)

predicting the future depend on rationalistic thinking: extrapolation of past trends into the future (Bradfield et al. 2005); this uses historical data, in turn, tested by econometric methods to find trends. However, one alternative to these methods is scenario-based planning. Scenarios play in the contractual environment (Van der Heijden 2010), their aim is to identify possible plans of action for say, air transport firms. Besides forecasting methods, scenarios are therefore one way to think about the future, in this case, of the airfreight sector.

Our purpose here is only to scan possible megatrends and driving forces that surround the airfreight sector and that are emerging or will emerge in the future, but we do not develop a scenario. A megatrend is a large economic, environmental, social and technological change that is slow to emerge.

The first rectangle of Fig. 4.8, on the left hand side, compares the driving forces of megatrends of airfreight. Figure 4.8 describes the contextual environment and the contractual environments surrounding the airfreight sector (including the logistics firms and stakeholders). This approach is useful in developing a typology of key factors that have emerged, or that will do so, and which will influence the future of the sector. These driving forces, found in the contextual environment, will shape the future of airfreight activity; whether it expands or contracts. For example, key forces (outside the sector) include declining/rising growth in high value exports and imports, declining world trade, rising fuel prices and vigorous business innovation (Fig. 4.8). Figure 4.8 can support our conceptual framework that is used to evaluate forces outside related activities of airfreight and of logistic firms.

It is also possible to discover discontinuities in the driving forces through the diagram (Fig. 4.8) if those forces rapidly change. What would happen if the forces in the left rectangle interacted with each other? i.e. new entrants (airports) appear and if the current financial crisis endures. These forces can shape the growth trajectory of airfreight transport. Firms can use the context (Fig. 4.8) to build scenarios to imagine the future of airfreight. We can summarize the various forces in the external environment of logistics organisations using the diagram in Fig. 4.8.

The contextual environment, of airfreight firms, can include the market forces as the background for planning future. The contextual environment also connects the megatrends with the decision-making process of the airfreight/logistics firms and the stakeholder. There are at least four megatrends: peak oil, competition from other world regions and demographic changes, and achieving a climate change agreement.

The contextual environment allows for the identification of the various mega-trends exerting pressure on the contractual environment of the airfreight firms. The forces of these trends are political, economic, regulatory and environmental (Zhang and Zhang 2002; Yamaguchi 2008).

In contrast to the contextual environment, the contractual environment is the field where firm managers or policy makers shake hands, this is encircled at the centre of the diagram. This field is led by the airfreight industry and its clients. They constitute the base of the airfreight market. The future growth trajectory of airfreight market (includes demand) and the management of logistics firms will be determined by the interaction between the two environments.

One can identify a driving force in the contractual case. How will aircraft industry and the air cargo industry react if changes in the driving forces of the megatrends occur? The growth of the rail freight mode will take market share from the airfreight mode; however, the latter has speed advantage, and these two modes compete for different markets. Growth in airfreight has taken place alongside the spread of high speed rail, but the latter should narrow the attractiveness of airfreight by reducing its speed advantage. In the contractual environment (in Fig. 4.8), one can identify two key actors: the freight integrators and the aircraft industry.

Figure 4.8 allows the reader to identify potential limiting factors for airfreight growth in future. This includes (1) the possibility of declining world trade, (2) achieving a global climate agreement which will impose a CO_2 emissions limit on the sector, forcing a downward shift of consumption of fossil energy of the air-freight sector; and (3) recurring financial crises/financial risk associated to not pricing in limits on CO_2 emissions in the capital expenditure decisions of airfreight corporations. Failing to acknowledge that not all fossil fuel will be available for moving goods by air transport, because of obligatory emission cuts, can alter stock market value of the sector and its financial position. Similarly, these factors will determine whether or not the expansion of airports in the European Union will continue. We have outlined briefly the key megatrends that will shape the future of the airfreight industry.

4.6 Conclusion

This chapter shows that the composition and volume of trade determines airfreight flows in the EU. The analysis of airports at the regional level shows that four regions will be decisive for the future of airfreight growth path. The study also illustrates the combined role of changes in the connectivity of airfreight flows and of population density. These changes will impact on the location of airports and on the volume of airfreight flows.

Airfreight activity has become a key sector for time critical products, despite high jet fuel prices; the sector is also important for high value products. Airfreight has grown enormously, in turn, generating the need for more airports, and in turn, increasing the need for better use of hub and spoke networks. Corporate actors of airfreight have adapted their services and their geography to compete for markets. These actors have modified the role and the position of airports. For example, urban airports do generate logistical activity which confirms the findings of Bel and Fageda (2005). These airports are being used as part of the niche strategies by some airlines. The airfreight sector has the ability to be the maker and the breaker of city economies repeating the process of road freight in shaping cities in the past.

As Kasarda and Lindsay (2011) have argued airfreight has the ability to alter the regional economies through its influence on the location of airports, logistics and manufacturing firms; this is the situation in North-West Europe where airfreight volumes and logistics firms' profits are found to be correlated using the evidence on 14 airports in Europe.

The above changes have multiple consequences for the future expansion of airfreight and for the environment-economy nexus. One limiting factor for the expansion of airfreight is the need to reduce fossil energy and related CO_2 emissions owing to the limited possibility of fuel switching in the sector. This means society will need to choose between the benefits given by fluid supply chains that depend on the speedy delivery of consumer goods and satisfying limits on CO_2 emissions.

We have identified five preliminary megatrends that will limit growth of airfreight in future. The most important of these is the rise of e-commerce combined with the need for speed by global supply chains; these two megatrends will favour this mode. These megatrends should be translated into actions for the individual airfreight firms. The preliminary analysis of megatrends surrounding the sector can be used to further develop scenarios of air transport for future research.

Future work should develop participatory scenarios to explain the future pathway of airfreight using the framework described here. The great recessions of 2008 and previous ones and the increases in jet fuel prices have not dented the growth of airfreight in Europe. However, solutions will be needed should climate agreements and decarbonisation be needed owing to events beyond the control of the sector.

Acknowledgements Dr. M. Strale (Université Libre de Bruxelles, Belgium) contributed towards this chapter with maps and literature sources. J. Navarro (UNAM, IIEC) edited the text and updated tables and graphs.

References

Ackoff RL (1974) Redesigning the future. Wiley, New York

Alkaabi KA (2011) The geography of air freight: connections to U.S. metropolitan economies. J Transp Geogr 19:1517–1529

Babikian R, Lucachko S, Waitz I (2002) The historical fuel efficiency characteristics of regional aircraft from technological, operational, and cost perspectives. J Air Transp Manag 8:289–400

Barker T, Dagoumas A, Rubin J (2009) The macroeconomic rebound effect and the world economy. Energy Effic 2:411–427

Bel G, Fageda X (2005) Getting there fast: globalization, intercontinental flights and location of headquarters. http://128.118.178.162/eps/urb/papers/0511/0511008.pdf. Accessed Jan 2014

Bocquet Y (2009) Les hubs de fret aérien express. Bulletin de l'Association des Géographes français 4:472–484

Boeing (2009) Fact sheet 747-8, Chicago. http://www.boeing.com/commercial/747family/747-8_fact_sheet.html. Accessed Jan 2015

Bowen J, Leinbach T (2004) Market concentration in the airfreight forwarding industry. Tijdschrift voor economische en sociale geografie 95(2):174–188

Bradfield R, Wright G, Burt G et al (2005) The origins and evolution of scenario techniques in long range business planning. Futures 37:795–812

Burghouwt G (2013) Airport capacity expansion strategies in the era of multi-hub networks. ITF/OECD discussion paper, roundtable: on airport capacity under constraints in large urban areas 21-23 Feb 2013

Button K, Yuan J (2012) Airfreight transport and economic development: an examination of causality. Urban Stud 50(2):329–340

Carbone V, Stone MA (2005) Growth and relational strategies used by the European logistics service providers: rationale and outcomes. Transp Res Part E 41:495–510

Carroué L (2002) Géographie de la mondialisation. Armand Colin, Paris

CIRIEC - Centre International de Recherches et d'Information sur l'Economie Publique, Sociale et Coopérative (2002) Evaluation des retombées économiques de l'aéroport de Liège-Bierset en termes d'emplois. Université de Liège, Belgium

Dicken P (2003) Global Shift: reshaping the global economic map in the 21st Century. Sage, New York

EC (2012) Reducing emissions from the aviation sector. https://ec.europa.eu/clima/policies/transport/aviation/index_en.htm. Accessed July 2014

Eurostat (2012a), EU transport in figures. Statistical Pocketbook 2012, Brussels. https://ec.europa.eu/transport/facts-fundings/statistics/pocketbook-2012_en. Accessed Dec 2014

Eurostat (2012b) Statistics, Brussels. http://epp.eurostat.ec.europa.eu/portal/page/portal/eurostat/home/. Accessed Dec 2014

Eurostat (2015) Eurostat regional yearbook 2015, Brussels. https://ec.europa.eu/eurostat/web/products-statistical-books/-/KS-HA-15-001. Accessed Dec 2016

Eurostat (2019) Globalisation patterns in EU trade and investment. Data: Extra EU Trade since 1999 by mode of transport (NSTR). https://appsso.eurostat.ec.europa.eu/nui/submitModified-Query.do. Accessed in March, 2019

Gardiner J, Ison S (2008) The geography of non-integrated cargo airlines: an international study. J Transp Geogr 16:55–62

Gardiner K, Ison S, Humphreys I (2005) Factor influencing cargo airlines' choice of airport: an international survey. J Air Transp Manag 11:393–399

Gilbert R, Perl A (2008) Transport revolutions: moving people and freight without oil. Earthscan, London

Hall RW (1989) Configuration of an overnight package air network. Transp Res Part A 23(2):139–149

Hesse M, Rodrigue J-P (2004) The transport geography of logistics and freight distribution. J Transp Geogr 12:171–184

Hummels D (2009) Globalisation and freight transport costs: maritime shipping and aviation. Challenges and opportunities in the downturn. Forum paper, 2009-3, ITF, Paris. http://www.internationaltransportforum.org/2009/workshops/pdf/Hummels.pdf. Accessed Feb 2016

Hwang C-C, Shiao G-C (2011) Analysing air cargo flows of international routes: an empirical study of Taiwan Taoyuan International airport. J Transp Geogr 19(4):738–744

IATA - International Air Transport Association (2013) A global approach to reducing aviation emissions. http://www.iata.org/publications/economics/public-policy/Pages/environment.aspx. Accessed Mar 2014

International Energy Agency (2012) Statistics, Paris. https://www.iea.org/statistics/index.html. Accessed Mar 2015

IPCC (1999) Aviation and the global atmosphere. Carbon intensity of airfreight. Factors for different major freight transport modes in Whitelegg, 1993; IPCC, 1996a; OECD, 1997a, downloadable at: http://www.grida.no/climate/ipcc/aviation/126.htm#img86. Figure 8-6. Special report of IPCC working groups I and II, 1999. Cambridge University Press. Accessed Mar 2016

IPCC (2012) Aviation and the global atmosphere. Chapter 8, II assessment report. http://www.grida.no/publications/other/ipcc_sr/?src=/climate/ipcc/aviation/126.htm. Accessed Apr 2015

Kasarda JD, Green JD (2005) Air cargo as an economic development engine: a note on opportunities and constraints. J Air Transp Manag 11:459–462

Kasarda JD, Lindsay G (2011) Aerotropolis: the way we live next. Penguin, London

Kiso F, Deljanin A (2009) Air freight and logistics services. Traffic Transp 21(4):291–298

Kohn C, Huge Brodin M. (2008) Centralised distribution systems and the environment: how increased transport work can decrease the environmental impact of logistics. International Journal of Logistics: Research and Applications 11(3):229–245

Lee J, Lukachko SP, Waitz IA et al (2011) Historical and future trends in aircraft performance, cost and emissions. Annu Rev Energy Environ 26:167–200

Leinbach T, Bowen J (2004) Air cargo services and the electronics industry in Southeast Asia. J Transp Geogr 4:299:321

McKinnon et al (2015) Performance measurement in freight transport: its contribution to the design, implementation and monitoring of public policy. International Transport Forum Report Paris 1–25. https://www.itf-oecd.org/sites/default/files/docs/mckinnon.pdf. Accessed Mar 2015

Negrey C, Osgood JL, Goetzke F (2011) One package at a time: the distributive world city. Int J Urban Reg Res 35(4):812–831

Noviello K, Cromley RG (1996) A comparison of the air passenger and air cargo industries with respect to hub location. Great Lake Geogr 3(2):75–85

Oum T, Xiaowen F, Zhang A (1999) Air transport liberalisation and its impacts on airline competition and air passenger traffic. International transport forum, OECD. Transport for a global economy: challenges and opportunities in the downturn. Paris. http://internationaltransportforum.org/jtrc//discussionPapers/DP201019.pdf. Accessed Sept 2016

Pels E (2008) The environmental impacts of international air transport: past trends and future perspectives. International transport forum. p 1–22, global forum on transport and environment in a Globalising World 10–12 November. Guadalajara, Mexico, p 2008

Pinder DA, Witherick ME (1972) The principles, practice and pitfalls of nearest neighbor analysis. Geography 57:277–278

Ramirez R (2008) Scenarios that provide clarity in addressing turbulence. In: Ramirez S, Van Der Heijden (eds) Business planning for turbulent times, 1st edn. Earthscan, London, pp 47–63

Reynolds-Feighan AJ (1994) The E.U. and U.S. air freight markets: network organization in a deregulated environment. Transp Rev 14(3):193–217

Royal Commission on Environmental Pollution (RCEP) (2002) The environmental effects of civil aircraft in flight. https://www.aef.org.uk/uploads/RCEP_Env__Effects_of_Aircraft__in__Flight_1.pdf. Accessed Nov 2015

Rutherford D, Zeinali M (2009) Efficiency trends for new commercial jet aircraft: 1960-2008, International Council on Clean Transportation. Report, pp 1-15, November 2009. https://www.theicct.org/sites/default/files/publications/ICCT_Aircraft_Efficiency_final.pdf. Accessed Apr 2014

Samii AK (2001) Stratégies logistiques. Fondements, Méthodes, Applications. Dunod, Paris

Strale (2010) La localisation des entreprises logistiques et le positionnement des régions urbaines nord-ouest européennes. Belgeo 1–2:119–134

Sun Y, Lu Q, Peng X, et al(2010) The role of logistics capabilities in promoting air cargo flows. In: 2nd international conference on information science and engineering, Hangzhou, 4-6/12/2010

Van Der Heijden K (2010) Scenarios: the art of strategic conversations. Wiley, Chichester

Wikipedia (2013a) On the definition of hub and spoke network. http://en.wikipedia.org/wiki/Spoke-hub_distribution_paradigm. Accessed June 2017

Wikipedia (2013b) Statistics on top ten air cargo airports in the world. http://en.wikipedia.org/wiki/World's_busiest_airports_by_cargo_traffic. Accessed June 2017

World Bank (2018) Air transport, Freight (million-ton-km), Washington DC. http://data.worldbank.org/indicator/IS.AIR.GOOD.MT.K1. Accessed Apr 2018

Yamaguchi K (2008) International trade and air cargo: analysis of US export and air transport policy. Transp Res Part E 44:653–663

Zhang A, Zhang Y (2002) Issues on liberalization of air cargo services in international aviation. J Air Transp Manag 8:275–287

Chapter 5
Airfreight Transport in China, Trade and the Airport Network

5.1 Motivation

Airfreight flows registered in China's airways are growing by 10% per annum in 2005–2015 (World Bank 2018) leading to greater pressure to add new infrastructure and to a bigger oil import bill. In this chapter, our goal is to first identify emerging trends shaping the sector and its determinants (trade, population growth, density, city size, type of export or export structures, among others) in the last decades. To do so, we use data sets for the airfreight transport network (ATNC) of 285 cities in China, at prefecture level, in order to understand the spatial distribution of airfreight sector using the city as the unit of analysis. Our second goal is to pinpoint the historical motivations of the recent history of that nation that influence the airfreight activity of 1975–2015. Airfreight transport refers to goods carried by aircraft. As Isard (1945) noted more than 70 years ago the aircraft can play a role in transporting perishables, high-grade commodities and irregular shipments as well allowing contact among key population centers.

The following six sections explain our diagnosis of air transport flows associated with the vigorous manufacturing industry and to strong export growth, both of which the freight market is composed.

Understanding the air cargo market is of paramount importance for three reasons: it can (a) guide the timing of investment decisions as well as (b) identify the geographical points for airport infrastructure including roads, bridges and warehouses and (c) identify future source of demand for jet fuel (air transport sector's energy use). China has 285 cities and it also has 20 cities in rank one mostly located in the eastern part of the country (National Bureau of Statistics China 2017), and these cities are served by 403 airports (World Bank 2009).

The airports make up the air network in China while its network structure represents transportation firms' production plans and the products that are offered. Different measures of network structure emphasize the aspects of concentration, dominance, connectivity, circuitry, duplication and centrality (Reynolds-Feingham

© Springer Nature Switzerland AG 2020   93
D. Bonilla, *Air Power and Freight*, SpringerBriefs in Energy,
https://doi.org/10.1007/978-3-030-27783-3_5

2013). All transport systems and their infrastructure depend on networks. Infrastructure is the term for all spatially connective investments and the associated rules and regulations. It includes roads and railways, airports and air transport systems, telecommunications and the Internet (World Bank 2009). Figure 5.1 shows the geography of China. The rail, road and water networks as well as the ATNC make up the total transport network of China.

China has 34 provincial led administrative units: 23 provinces, 4 municipalities (Beijing, Tianjin, Shanghai and Chongqing) and 5 autonomous regions (Guanxi, inner Mongolia, Tibet, Ningxia and Xinjiang).

The administrative units along the coast, the North-West and South-East are largely covered by the road network (Fig. 5.2).

The road network is highly concentrated to the coast and the south of China. The North-West region has been largely neglected by successive administrations and not

Fig. 5.1 Map of China's provinces. (Source: http://www.orangesmile.com/common/img_country_maps/china-map-0.jpg)

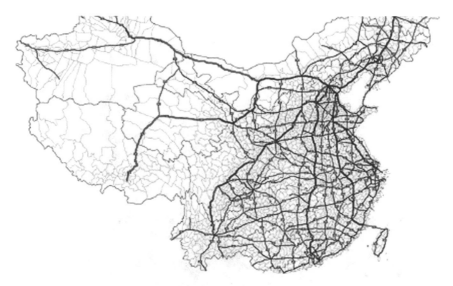

Fig. 5.2 National trunk highway systems in the People's Republic of China. (Source: http://www.roadtraffic-technology.com/projects/national-trunk-highway-system/)

all regions show network development. The map (Fig. 5.2) shows that the Western and Central regions have a lower road density than the Eastern regions do; the regions less well served also display a lower population density.

The ATNC mirrors the shape of the road network: the density of the air transport network is highest in the south Eastern region while the Western region lacks network development. The North-West region shows an extensive network. Figure 5.3 describes the ATNC of China.

The advantage of the ATNC can complement the weaknesses of the road network. We can speculate that the road network relies on the ATNC and vice versa: the ATNC feeds cargo volume by the road network and the ATNC also feeds cargo volume into the road network.

A formidable competitor for the ATNC network is the newly built high-speed rail network (HSRN; Fig. 5.4a) which bears potential for high-speed freight. Both planned and current networks are shown (Fig. Fig. 5.4a, b). The rail mode can move freight at high speed within China's transport networks; the speeds that HSRN can achieve can get closer to that of the aircraft, cutting off the advantage of the latter. Although at the moment the HSRN serves the passenger sector, in future the freight market can also be served by the HSRN.

Naturally, as the rail (particularly HSRN) and road networks are further deployed in the territory, they potentially can reduce demand for airfreight flows. The opposite effect can also occur: HSRN can stimulate demand for airfreight flows.

The goal of the Chinese Government is to expand the HSRN track to 45,000 km of length, which is more than enough to encircle the world (StraitsTimes 2017), 29 provinces are connected to the HSRN. The HSRN can achieve a speed of 350 km/h (Ibid.)

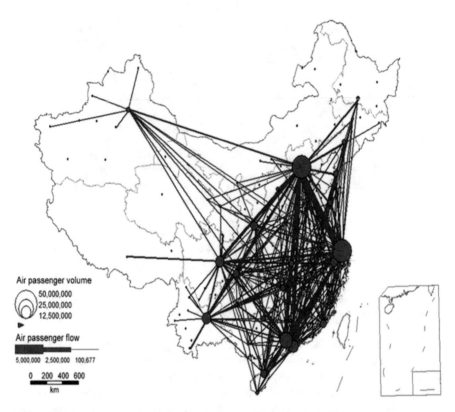

Fig. 5.3 Air transport network for key cities. (Source: Wang and Jin 2007)

As far as density of network, the HSRN of China resembles the geographical spread of the ATNC network: The Eastern region is highly developed with more rail links than the western region. The North-West region shows some network development; however, the western and central regions show far less network development or none at all. In short the development of the HSRN is highly biased for the coast of the country as the rest of the networks do. The HSRN can also complement the ATNC; however, in some cases the HSRN can outcompete the ATNC as far as speed and reliability.

5.2   Qualitative Factors and the Expansion of Airfreight Shipments

We discuss in this section the qualitative factors of airfreight growth. These and other factors are discussed in detail in Sect. 5.4. First is the modal split which will vary among regions because of the different geography and because each transport mode serves a different function within China's geography: The railways will have

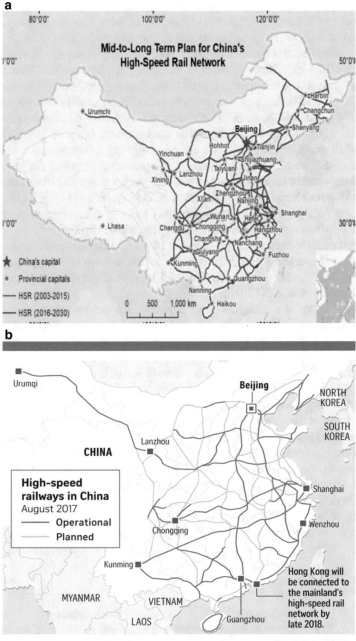

Fig. 5.4 (**a**) China's current and planned HSR network. (Source: https://www.straitstimes.com/asia/chinas-rail-ambitions-run-at-full-speed. 2017, Sept.). (**b**) High-speed railways in China. (Source: Strait Times: https://www.straitstimes.com/asia/chinas-rail-ambitions-run-at-full-speed. 2017, Sept.)

an advantage over air transport mode in some cases (this mode can carry a large volume of low value product), but the air mode will have speedier delivery systems whilst in some regions the road mode will be the preferred option.

The second factor is the socio-technical regime along with path dependence and technology lock-in errors (Banister et al. 2011) that has set in China's freight transport system. The technology regime is the "inertia of established technologies" (Geels 2002) also defined as "the socio technical configuration of transportation" (Geels 2002; Banister et al. 2011). Configuration refers to a heterogeneous set of elements, whose addition indicates that configuration fulfils a function (Geels 2002). Path dependence refers to a minor or fleeting advantage or a seemingly inconsequential lead for some technology, and product or standard can have important and irreversible influences on the ultimate market allocation of resources even in a world characterized by voluntary decision (Liebowitz and Margolis 1995, p. 1). Technology lock-in refers to an economy that locks-in to incorrect choices i.e. jet engines, vehicles, nuclear power stations and keyboards (Ibid).

A third development is the government target of airfreight expansion. China's airfreight movements are predicted to reach 170 billion t-km in 2020 with an annual growth of 12.2%, and aviation services will by then cover 89% of its population (13th Five Year Plan, CPC 2016).

A fourth factor of airfreight activity is subdivided into two economy wide factors: (a) large-scale investment in airport facilities and (b) transport infrastructure that has taken China from a small air cargo market to one of the largest air cargo hubs in the world.

A further factor related to the above points (a and b) is the growth of trade which lifts that of air cargo, and the latter is positively correlated to economy wide performance. Air cargo growth is twice that of GDP for most of the period considered. Table 5.1 shows the relationship between air cargo and GDP growth that has held for decades but breaks down in the 2005–2015 period after the global financial crisis of 2008 that triggered a decline in global trade. Prior to that period (1975–2005), air cargo growth overtakes that of GDP. China's growth rates in air cargo activity (Table 5.1) supersede those of Japan and the USA in earlier phases.

A fifth driver of freight flows is investment. To match the expected demand in airfreight volume and the associated demand for exports and imports for transport services, investment in airports will need to be accelerated. New regional airports and larger hubs need to be planned, while the existing airport capacities will need to

Table 5.1 Average annual growth of GDP and air cargo: China

Year	GDP (current US$ 2015) Compound annual growth rates in %	Air cargo (tonnage) Compound annual growth rates
1975–1985	6.67	24.18
1985–1995	9.06	14.05
1995–2005	11.98	17.58
2005–2015	16.96	10.08

Source: World Bank (2018) (in %)

be updated. All of these airports can lead to regional economic development but also to greater inequality among regions. The airport links the economy of the local area and the world through passengers and freight movements, but the basic link between the airport and the economy is trade at both regional and international levels, increasing the consumption of goods and services in urban areas and local production activities. The city provides these sources of economic growth and air transport development because the city is the centre of trade at regional level. But this was not always so. In the 1950s–1970s period, the rail mode was the main mode, during which the domestic market growth was the main policy priority rather than the off-shore export market (Table 5.2).

The commercial success of cities requires investment, import substitution and exports. The latter need airports. Two hundred airports are added every year (National Bureau of Statistics China 2017) and besides building hundreds of new airports to support the airfreight industry, China is building roads, bridges, power plants and rail track infrastructure that will complement the growth of the airfreight sector.

In Table 5.2 China's modal split is tabulated (freight combined). One can see the freight traffic is dominated by the road mode in recent years; this is the most competitive mode. The rapid increase in both road and aviation modes took place in the 1970s. Competition between the rail and air modes took place in the 1990s when the air mode showed a much larger growth rate than that of the rail mode, albeit from a lower base level for the air mode. Competition will continue among the air, the rail and road modes. What led to these changes in modal split? The first answer is the large-scale infrastructure investment that has been planned as part of China's transport policy that continues to favour the air and road modes, the second answer is the low value of transport fuels (diesel, jet fuel, fuel oil and crude oil) in most of the period under observation.

Table 5.2 Freight traffic by mode (million tonnes), 1952–2015

	Total	Railway	Highway	Waterways	Aviation	Pipelines
1952	315.16	132.17	132	51.41	0.002	0
1962	803.62	352.61	328	254.44	0.018	0
1970	1503.59	681.32	568	426.76	0.037	0
1980	5465.37	1112.79	3820	468.33	0.089	105.2
1990	9706.02	1506.81	7240	800.94	0.37	157.5
2000	13,586.82	1785.81	10,388	1223.91	1.967	187.0
2010	32,418.07	3642.71	24,481	3789.49	5.63	499.72
2015	41,758.86	3358.01	31,500	6135.67	6.293	758.7

Since 2000 data includes the ocean freight mode. Since 1979 the freight traffic by highways has included quantities transported by trucks of non-highway departments. Since 1984 it has also included quantities transported by private trucks
Rail includes national and local railway traffic
Source: National Bureau of Statistics China (2016c). Source: http://www.stats.gov.cn/english/statisticaldata/yearlydata/YB2000e/O08E.html

In the 1950s the rail mode is the key transport mode; in the 1960s this pattern continues; however, by the 1970s the road mode was given priority by the transport ministry of China. By the 1980s, it appears that the rail mode is overtaken by the road mode suddenly. Since then the dominance of the latter continues unabated (Table 5.2). The waterway mode does lag significantly behind the other modes; however, this mode is equally important.

5.3   Literature Review of Airfreight Flow and Networks

The review covers the literatures of (1) time space, (2) social commentary, (3) urban theory, (4) transport network theory and (5) transport and development, and more substantially of (6) economic geography which helps explain the causal link between wider economic growth, the latter's benefits and that of the airfreight transport. Broadly speaking the objective of this review is to connect the different strands of literatures of various fields to explain the emergence at the regional level of airfreight shipments. The review is by no means complete.

5.3.1   Freight, Time and Space

Space (mainly distance) and time are the basic tenets of air transport movements, and speed is key to the advantage of the airfreight mode. Airfreight flows are transforming the way humanity experiences space and time. Time is a strong feature of the airfreight mode: in 1930–1960 aircraft speed has achieved 100–300 knots (IPCC 1999). Mach speed is now achieved by a 787 aircraft of 0.85, and since 1960 that speed remains the standard for commercial aircraft. The majority of turbofan-powered aircrafts in today's world fleet have average cruise speeds of about 500 knots (IPCC 1999, Special Report on Aviation).

Time drives modern mobility of goods and people. Galileo first changed our perception of time after the invention of the clock (reference needed). This was possible because of the cartography of time. Cartography refers to the science or practice of drawing maps. Cartography, the compass and the clock changed transportation networks for ever: The compass is an instrument used to navigate and for orienting shippers, and it shows direction using cardinal coordinates or points.

Marx (1857) treats time as the annihilation of distance from the viewpoint of an economist 'capital by its very nature drives beyond every spatial barrier. Thus the creation of the new means of communication and transport, the annihilation of space and time, becomes and extraordinary necessity for it'. In the latter's view, the pressures of the capitalist system drives the bourgeoisie 'to nestle everywhere, settle everywhere and establish connections everywhere' (Stanford Encyclopedia of Philosophy or S.E.P. 2014; Scheuerman 2014).

To establish connections everywhere, transport infrastructure development is needed which itself depends on attaining speed. High-speed modes of transport and the Internet represent new challenges for citizen participation (McLuhan 1964, p. 103; Dewey 1927; Virilio 1977), and speedier modes of transport strengthen the executive branch at the expense of the legislative one (Virilio 1977). Speed is considered as part of the war on time (Virilio 1977). In sum modern transport relies on the technology developments that affect speed levels of aircraft.

Literature, philosophy and social commentary have appeared on the subject of time, transport activity and speed. The appearance of high-speed forms of human activity generated much writing about the compression of space (S.E.P. 2014). Nineteenth- and twentieth-century literature contains numerous references about the transformative power of high-speed forms of transportation (rail, air and car) and communications (the telegraph and telephone) which opens opportunities for greater spatial interaction across geographic and political dimensions. Heine's words (cited in Schivelbusch 1978, p. 34) argues that 'space is killed by the railways and we are left with time alone'. While Auden (1936) observes the spatial effect of the railways when describing freight deliveries as 'This is the night mail crossing the border, bringing the cheque and the postal order' (S.E.P. 2002).

Numerous literary works describe vividly how time and space (distance) are transformed by new transport modes and how these alter the travel experience. The mail delivered (described in Auden 1936) takes place in a dedicated freight train moving from Euston station in London to Scotland Glasgow and Edinburgh and to Aberdeen (Britain). The poem reveals three features about freight delivery: (1) the mail system is complex and it requires National Government in order to succeed, (2) it is a symbol of modern efficiency and (3) mailmen are industrious and jovial (McLane 2012). The rail mode is better suited, than the airfreight mode at the time, for mail delivery: picking up, setting down and sorting mail while travelling. The poem of Auden also reveals (1) the emergence of the age of information flows through mail and rail transport and (2) on how humans experience time through work activity; this experience is rapidly evolving in the present century.

In Orlando early in the twentieth century the concept of the elasticity of time is developed: 'why our perception of time differs as we age, slows down when under pressure and time stops when humans go on vacation' (described in Woolf 1928).

Instead of developing further the concept of the elasticity of time, some authors describe time as one of flows of information and of goods (Castells 1996). For this writer space and time make up the network society. Previous forms are space of flows and timeless time. The former refers to the social interaction in chosen time at a distance through networking communication with the support of fast transport technologies; and the latter refers to the desequencing of social action by time compression or by the random ordering of the moments of the sequence (Castells 2007). Space and time are related in nature and in society (Castells 1996). In this case space structures time. If time is producing flows of information, greater volumes of these must have impacts on airfreight activity.

Time and space are altered in the urban space. It has been argued that as distance was annihilated "distance, the surface of our country would shrivel in size until it became not much bigger than one immense city" (Harvey 1996). Space or distance is then hugely important for transport (Banister 2010). Following the steps of Heidegger who first observes the abolition of distance as a key feature of contemporary life, an economist coins the term the "death of distance" (Cairncross 1997). That author builds on Marx's concepts on space and time. However, a casual inspection of global trends shows that distance travelled continues to increase in many nations and in China. In relative terms the distances in the twenty-first century are getting shorter on average passenger trips, but the number of these trips per capita continues to increase. The freight transport sector (road, air and railways) continues to record higher average distance hauled in many countries.

Others scholars have been concerned with the annihilation of space or time. The possibility of no space and no-time elements appears in the 'Circular Ruins' (Borges 1944). Although for Borges time travel occurs on a mythical level and not through machines, the writer borrows from the optic of geography and so of space to explain the concept of time.

Borges describes the features of time as:

'Every instant is autonomous... No less vain to my mind are hope and fear, for they always refer to future events, that is, to events which will not happen to us, who are the diminutive present. They tell me that the present, the "specious present" of the psychologists, lasts between several seconds and the smallest fraction of a second, which is also how long the history of the universe lasts...'. In the above paragraph Borges sustains that language is dominated by the topic of time. The theoretical implication of this concept of time is that freight transport is heavily dependent on ever decreasing fractions of time, on an autonomous instant and on a diminutive present.

Borges says that time is the succession of ideas (Quoting Berkeley, Principles of Human Knowledge, p. 98). And he further writes 'once matter and spirit, are negated, once space too is negated I do not know with what right we retain that continuity which is time'. In other words he explores the idea of the rejection of time.

Changes across time, distance, speed and higher average distance: these are symbols of our times. How is this relevant for airfreight flows? If we were to negate space we would reject the need to move speedily both goods and people to match time budgets. Borges places time as central to human identity in his literature. That identity can be built also by the movement of goods or freight transport.

Time is a key success factor of a modern freight transport industry. Time savings is also a key element of the new managerial technique of 'Just-in-Time' (JIT) delivery. The technique favours the airfreight mode since the mode saves time for delivery of goods. The prime goal of JIT is the achievement of zero inventory, not just within the confines of single organisation, but ultimately throughout the entire supply chain (Hutchins 1999). The dynamism and growth of the airfreight market depend on how quickly exports and imports are moved within and among countries (Nordas 2007). Therefore JIT practices for airfreight management is a key part of this.

In short, how society views, understands and experiences time will determine the evolution of airfreight flows through the demand for quick delivery of goods. The dynamism of the airfreight transport sector can also be explained from the optic of urban theory and urban economics. Within the latter framework network theory also contributes to understanding the links among urban freight movement, the airfreight market and the system of cities which trade with each other commodities and high-value products. The system of cities is the backbone of air flows and trade within and between regions. Besides literature and social commentary, network theory affords a new dimension to explaining the role of space, time and airfreight flows.

5.3.2   Networks and Airfreight

Cities, freight transport movements among cities, regions and countries, and transport networks are connected through roads, bridges, airports, rail stations, the Internet and e-commerce. Economic life is surrounded by networks. A network is composed by links which give shape to a system. A transport network is made up of nodes and links that allow the movement of goods and people (Ducruet 2017). A node is a city or an airport, and a link (also an edge) can be the number of routes of airfreight flows. The theory of networks teaches that airfreight networks reinforce each other.

A network can be explained by graph theory (Acatitla and Alonso 2017). Graphs are models that represent topological features which are essential for a network. For example in the air transport network, an airport or a city represents the nodes, and the routes of air cargo flows are represented by the edges.

Airports and air flows (airfreight activity is one element of it) provide the back-bone of world city networks as described in Choi et al. (2006) and Mahutga et al. (2010), besides these two authors there is little work on the relationship between transport and regional economic development from the viewpoint of city networks (Lao et al. 2016). A city network is defined as an interlocking network (Pereira and Derudder 2010). Following Taylor (2001) there are three levels: the network, nodes (cities) and sub-modes (global service firms). The city network can be connected by the airport network, the rail network, the road one and the Internet.

The theory of modern transport networks is first developed by Kansky (1963) who calculates three indexes: alpha, beta and gamma to identify the type of network. Using Kansky's indexes, Stokes (1968) calculates these indexes for the freight transport of Venezuela and Colombia (rail, road, air and maritime modes). The Stokes study is one of the first ones for a developing country. These researchers rely on scant data sets and on basic computational ability which limits their analysis of networks. Relying on larger data sets on transport networks, some researchers (Jin et al. 2010; Ducruet 2017) have devised quantitative measures which superimposes the strength of several networks to explain the success of the transport network (including the air transport system) of the ATNC.

Pioneers of city network research include Castells (1996) and Camagni (1993), but these authors avoid analysis of transport networks with a special reference to airfreight flows. Barabasi and Albert (1999) take the field one step ahead by examining the network features by means of mathematical models but at the expense of economic and geography principles underpinning networks. Barabasi and Albert

(1999) find that networks can have a varying distribution of links and are not always random networks as Erdos and Renyi (1960) first argued. The availability of data and new quantitative tools for analyses allows theorists to improve the theory of networks; these tools show that the classical theory of random networks lacks realism. Networks are complex, dynamic and can have new nodes (more cities) and edges (air cargo flows).

Network theory can be applied to airfreight flows (airline markets) as well as to systems of cities. Wang et al. (2011, 2014) and Lin (2012) apply network theory to deregulation of the air flows; these authors develop measures of network structure of air transport flows of the ATNC.

Another feature of the city network research is the economic network method (of which airfreight activity is part) that can be measured from the lens of corporate organizations, and inter-firm relationships as in Lao et al. (2016) and Taylor et al. (2002). This strand of work focuses on the correlation between the corporate network and that of the city on a global scale.

5.3.3   Economic Development and Airfreight Flows

The shape of the economic network can be influenced by the economic development of cities and of nations which is closely related to transport infrastructure (Banister and Berechman 2000; Hickman et al. 2015) and thereby to freight transport. The causal relationship between development and transport investment, however, can be undetermined: economic development generates demand for transport infrastructure, and new additions to infrastructure, in turn, create economic change. The phenomenon is widely explored in the economic geography literature.

In the middle of the twentieth century, Taaffe et al. (1960) explores the link between transport and underdevelopment and concludes that population, the physical environment, rail competition, intermediate location and commercialization are motivations for transport investment.

In the same vein, after observing the economic implications of aircraft, it has been found that these will affect (1) population movement and (2) the structural and functional relations between city and country (Isard 1945). These two effects will create opportunities for transport investment, say for airports, whose effects are mainly a reduction of freight costs and shorter travel times and an increase in exports.

The impacts of aircraft, which include reductions in freight rates, resemble those of the street, the electric railway, the automobile, the railroad and canal (produced by reductions in the cost of freight movement). Inherent in air cargo flows are the emergence of new industrial and commercial areas and population agglomerations, alterations of trade channels, and outlets for new investment (Isard 1945, p. 164).

In contrast to Isard (1945) who measures the system wide benefits of aircraft, Katz and Shapiro (1994) explore competition among technology systems (including transport systems, its components, repair parts and services) and network effects.

These effects can arise when users join a network and adoption externalities can also result.

More recently, the analysis of the New Economic geography (NEG) optic complements the theoretical body on the network effects of aircraft. The NEG approach explains the location of manufacturing industry (which determines distance travelled by a truck or hauling distance) which also requires freight transport services (for inputs of raw material), including airfreight flows. The location, or geographical distribution, of manufacturing establishments in urban regions is key for exports and imports. These activities need transport networks. The NEG posits that the growth of the urban economy stimulates the manufacturing sector. The NEG focuses on increasing returns to scale (an increase in all inputs i.e. labour and capital leads to a more than a proportional increase in output) and transport costs both of which are key influence factors of the economic and manufacturing success of cities. Increasing returns lead to agglomeration economies which lead to unequal growth among regions and cities whilst the latter have the advantage of agglomeration effects. Agglomeration effects can explain the advantages of clustering industrial and manufacturing activities, which in turn gives rise to the concentration of airfreight transport flows in a few city regions or airports. These effects require locating in areas heavily populated by other producers (IBRD/World Bank 2009, p.126).

In the NEG the spatial agglomeration of economic activity in a major core urban region is argued to have a positive effect on national growth as a whole (Clarke et al. 2016). Trade and urban growth can also influence national growth (Krugman & Livas 1996).

Krugman (1998) emphasizes distance as a key factor in the NEG, but it ignores density. Distance and transport cost are key variables for the NEG, but for economic geographers density, diversity and intensity are key variables that explain factory location and agglomeration (Parnreiter 2018, p. 109). Distance and cost are not sufficient to explain the effects of agglomeration. Factory location, which influences the clustering of economic activity, will be determined by both centripetal and centrifugal forces (Krugman 1998). Centripetal forces (which concentrate manufacturing activity) and centrifugal ones (which disperse economic activity) explain the clustering of economic activity. These forces can be market size or pure external economies or airfreight transport. In addition these same centripetal forces indirectly explain the clustering, or lack of dispersion of the sources of freight transport flows.

Unlike Krugman (1998) in the framework of General Equilibrium Theory or "GET" (Walras 1874) distance which relates to the spatial agglomeration is missing to understand the economic growth of cities. According to GET city, economies face constant returns to scale, perfect competition, no externalities, equal consumer preferences, flexible prices and flexible markets; however, the theory is able to explain neither agglomeration economies nor the concentration of economic activity.

The production function approach which is a by-product of the GET can be used to measure the effect of transport investment on regional economic development, this effect has been studied in many countries based on econometric and on input output models. In the 1990s, a renewed interest led several authors to propose a model to explain the effect of public infrastructure on national production, firing off an important policy debate on infrastructure (Aschauer 1990). These production functions can serve to isolate key variables of interest, but functions need to be complemented by interviews (Rietveld 1989). These methodologies have been influential in government departments, many of whom do publish forecasts using the production function approach in Europe, the USA and Japan.

A case in point is the large-scale review of transport infrastructure studies (UK DfT 2005) on the wider economic impacts which recommend the use of spatial general equilibrium models and regional economic models but asserts that the former models is still in its infancy (UK, DfT 2005).

The review presented above discusses theory that can help explain the growth of air transport networks outside the sector itself. The literatures on freight, time and space have been discussed in this section with particular emphasis on (1) the NEG, (2) how networks of cities affect the need to move freight among them and (3) how the externalities of agglomeration and of networks operate. The literature review presented in this section opens two lines of research for the analysis of air transport. The first is the overall effect of the time compression requirement on the modern economy as a whole, and the second is the effect on policy making. We now describe each of these.

1. Aircraft speed, as a time-saving device will continue to improve; this practice will continue to spread so long as the world economy is deeply tied to profit seeking behaviour enabled by management practices based on JIT.
2. Changes in decision-making will be enabled through the ongoing compression of time and of space that accompany the increased power of the executive branch of States around the world at the expense of other powers i.e. the legislative one, as Virilio (1977) has argued long ago. In other words, aircraft cargo can be a tool to strengthen the executive branch and thereby of decision-making.

In the following empirical section of the chapter, we attempt to employ network theory. Data limitations do not permit a full quantitative analysis of network analysis at the city system level using freight transport flows as the metric of analysis; however, this is a first step to using such network analysis of airfreight activity among cities.

5.4   Historical Developments of China's Airfreight Activity

Several historical events that shape policy making explain the evolution of airfreight shipments which have grown much faster than many trade indicators (Table 5.2), indicating a high elasticity of airfreight flows to export growth. Table 5.2 in Sect. 5.2 tabulates the historical evolution of the airfreight mode for 1950s through 2015. One of the most important policy developments that shaped the air cargo industry is the deregulation of the air transport sector in (Wang et al. 2016) in the 1990s which seeks to increase domestic competition. In 1949–1978 China's civil airports were closed to international aviation and were run on non-commercial grounds which inhibited the growth of the air cargo sector. In that period China's economy operated under a system of import substitution which may have strengthened the influence of a few large cities while weakening the hinterland. During 1950–1978 the economy is geared towards the domestic market which may have favoured the railways rather than airfreight flows.

In the export-led era starting in 1978 (exports accelerate in 2000 after the WTO entry), the importance of new centres of industrial and economic power grew at the expense of the hinterland. These changes shaped the dynamism and likely raised the volume of airfreight flows.

Beside changes in the market regulation of air transport, other economic motivations that play a role for the development of airfreight include:

- Urbanization
- Population density
- Transportation costs
- City size and regional trade
- Wages
- High industrial activity
- Mobile phone diffusion
- Intermodal competition

We now discuss each of these factors. Table 5.3 shows that since 1990–2016, China's urbanization rate is slowing down; however, in absolute terms this is already a high level of urbanization because of its large population (upwards one billion people).

The density of population (population mass per square land area) and of economic activity will vary by region. Population density reflects market size of airfreight-bound goods and labour availability in the industry.

Table 5.3 Key urbanization indicators for China

	1990	2000	2010	2016	AAGR
Urban pop.					
Growth (annual %)	4.31	3.65	3.26	2.61	
Airfreight (billion tonne-km)	0.8	5.27	17.89	20.8	9.4
Pop. density (people per square kilometre of land area)	120.92	134.49	142.49	146.85	0.5
Mobile cellular subscriptions (per 100 people)	0.00	6.66	63.17	96.88	21.0
High-technology exports (% of manufactured exports)	..	18.98	27.51	..	N.A
Services, value added (% of GDP)	32.38	39.79	44.07	51.63	1.30
Exports of goods and services (% of GDP)	14.03	21.24	26.27	19.64	0.94
Imports of goods and services (% of GDP)	10.66	18.52	22.62	17.42	1.39

World Bank (2018). World Development Indicators

A quick glance at official statistics shows that China's urbanization effort is one of the largest, and fastest, in the world (World Bank 2018), and this effort relies on road (road kilometre per 100 km), rail (rail kilometre per 100 km) and air transport networks. The density of these networks lies below comparable countries which indicates that China's urbanization push has plenty of slack for growth. A revealing indicator for further growth is that only 18% of China's population lives in cities above one million people compared to Canada's 45%, the USA's 43% and Japan's 48% (World Development Report, UN 2009).

Rapid urbanization is reflected in the number of newly built cities in the country. China has built more cities (about 125–150 cities) larger than one million people than the EU (36 cities) and the USA (9 cities) (Kasarda and Lindsay 2011). These cities increasingly produce, consume and import more goods all of which requires airfreight services. Therefore urbanization changes are key to explain the demand for airfreight flows regionally.

Urbanization is heavily influenced by population density which has increased considerably in the same period, but it is still considerably low (147 people per land area), which is below Japan (351) but above the EU's level (116) (data from the World Bank 2009). China's service economy has increased below the rate of urbanization, but the former has grown above the rate of mobile phone ownership. All of these indicators translate into annual growth of 1.9% of high-tech exports in the same period.

A third push factor is declining airfreight costs. These costs are an incentive for airfreight expansion and are heavily influenced by the cost of jet fuel, and this cost has fallen rapidly for two reasons: lower energy prices and engine efficiency gains. Jet engines are faster, more fuel efficient and reliable and require much less maintenance compared to the piston engines they replaced (Hummels 2007).

A fourth contributing factor is city size. In Fig. 5.5 we can see that city size, a product of urbanization, also determines freight volumes. The first 20 cities move most of the cargo (all modes) while small cities move very little air cargo.

Shanghai handles 6000 times more airfreight volume than China's smallest city in our sample. The largest 50 cities move almost half of all airfreight volume. Unlike the first top cities, ten cities move 10% of all air cargo. Air cargo flows are clustered in China's top 20 cities, mostly in the eastern region, taking 80% of total air cargo in 2014. Since the 1980s three Eastern cities (Beijing, Shanghai, Guangzhou) have taken the largest volume of air cargo, about 50% of China's air cargo. Cities with more than one airport include Shanghai, Beijing, Chengdou, Hangzhou, Nanjing, Guilin, Fuzhou, Jinan and Yinchuan.

A fifth contributing factor of airfreight flows is the type of economic activity affecting trade in a region, whether agriculture, heavy industry, or high technology. See below for a further discussion on trade factors for airfreight flows. The latter sector requires just-in-time deliveries which favours the air transport mode. Further influences on the airfreight flows include world economic activity, impact of services in the express and small package markets, changes in inventory techniques, deregulation and liberalization (discussed above), national development programmes and a stream of new air eligible commodities (Reynolds-Feigham 2013).

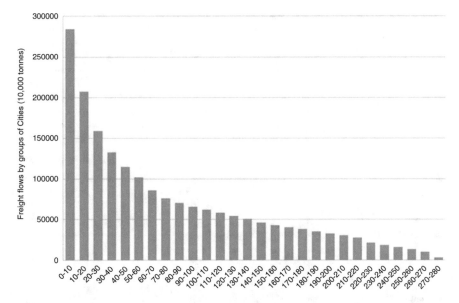

Fig. 5.5 City groups classified by the largest volume of freight by smallest volume of freight (volume of freight "Y" axis (in units of 10,000 tonnes)). (Source: China City Statistical Yearbook (2015))

Table 5.3 shows key indicators of urbanization that stimulate demand for airfreight transport.

A sixth factor is the newly emerging motivation for airfreight expansion which is the convenience of using just a click to order airfreight deliveries via mobile phone applications that are widely available to the Chinese consumer. Mobile subscriptions have almost achieved market saturation (Table 5.3). It is yet unknown how e-commerce via mobile applications will expand the volume of the airfreight flows, but it is likely the trend will slope upwards in the future.

Additional factors include regional trade (discussed below), wages and high-tech industrial activity (Lakew and Tok 2015); in the following section, we discuss the role of trade in determining China's airfreight volume.

5.4.1   China's Trade and Airfreight Activity

Economists largely agree that trade is the key driver of growth in airfreight movements (Hummels 2007). Figure 5.6 shows freight moved on the "Y" axis and the value of trade in the "Z" axis.

Figure 5.6 shows a positive relationship between airfreight activity (in t-km) and China's exports from 1980 to 2012 (in $ value). Airfreight takes a small share of the exports by the weight measure, but the relationship is also positive. Soaring exports

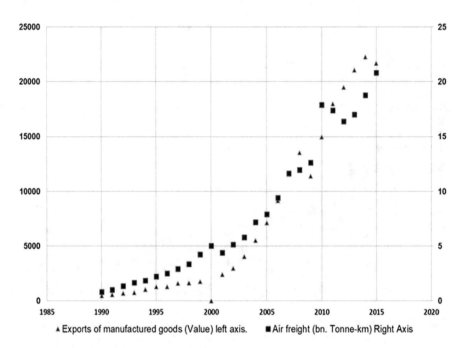

Fig. 5.6 Manufactured Exports (100 million US$) on the "Y" axis (100 million US$, 2015) and airfreight flows on the "Z" axis (billion t-km). (Source: National Bureau of Statistics China (2016a, b, c, d))

raise the demand of airfreight, while the airport guarantees and promotes trade activity.

In the eleventh 5-year plan, China announced that it would build a hundred new airports by 2020 valued at 62 billion US$ (Kasarda and Lindsay 2011, p. 387). In the 13th 5-year plan China then announced 260 new airports (CPC 2016, China five year plan).

China's *National Development Reform Commission* outlines the plan for the reform and development of the Pearl Delta river in 2008–2010. Thirty years ago China carried less than a billion t-km of air cargo (Fig. 5.6), on a par with Spain and below Singapore's level (World Bank 2018). In 2015 it carried 20 billion t-km of cargo (Fig. 5.6 right axis). China's push for aviation is correlated to its wish to stimulate its export economy. The country is also manufacturing its own aircraft to move air cargo.

In terms of value, air transport is an important service trade. China–US trade in 2000 via air cargo is valued at 19 billion US$ (2016 exchange rates), which takes 17% of China–US trade in the same year (U.S: Census Bureau 2017). By 2017, air cargo takes 30% of total trade between the two nations with the net balance in China's favour. China–US air cargo has almost doubled between 2006 and 2017 to 153 billion US$ (2017 US$) (U.S. Census Bureau 2017). The air cargo mode has grown less quickly than the shipping mode (Fig. 5.7) although the former carries high-value goods of US–China trade.

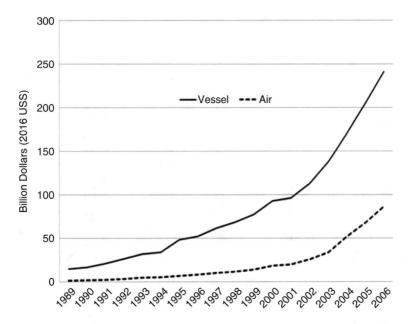

Fig. 5.7 US–China trade by air. (Source: FT 920 U.S. Merchandise Trade: Highlights, U.S. Census Bureau, 2017, various years)

Figure 5.7 shows that trade via air transport has grown quickly between China and the USA particularly after China's entry to the WTO in 2001. The US exports by the air cargo mode have grown less rapidly than those of China (not shown in the graph) (based on US Census Bureau Data 2017). China–US air cargo flows have grown by 4200% in 1989–2006, but there is a large trade gap between China and the US, amounting to 371 billion US$ in 2017 (Congressional Research Service 2018).

Although China is the biggest exporter, its products are mostly low-value products, so that they need fewer air transport services explaining the lower share of air cargo in total trade. The share of the shipping mode in total trade dominates the China–US trade, but the share has fallen in the last few years. It should be noted that since the China–US trade relationship is the largest of all trading activities of China, the relationship reveals the strengths of these modes.

5.4.2   EU–China Export and Airfreight Flows

The European Union (EU) was for a while almost the biggest export partner of China in 2007–2017, but the USA has now become the largest export market for China's products. According to China's official statistics, China imports more or less the same volume in value from the EU as from the USA with Japan trailing behind, although the recent past shows the USA is superseding the EU as the prime export market for China (Fig. 5.8).

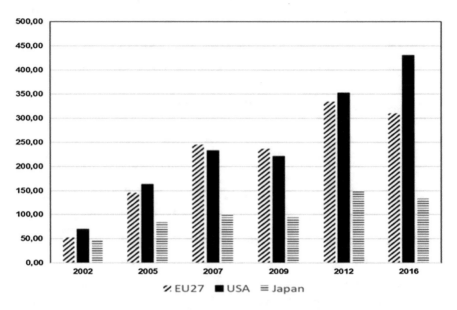

Fig. 5.8 China's key export markets and trade partners. (Source: World Integrated Trade solution, World Bank (2019))

Table 5.4 China–EU trade by transport mode

Year	Exports (billion euro, 2014)	Exports by air (%)	Exports by sea (%)	Export (million tonnes)	Other (%)	By air (%)	By sea (%)
1999	52.60	13.49	63.59	18.84	20.86	1.18	77.96
2000	74.63	14.96	61.52	22.48	20.56	1.20	78.24
2001	82.00	13.49	60.62	27.09	23.87	1.12	75.01
2002	90.15	13.78	58.11	26.02	21.65	1.31	77.04
2003	106.22	15.98	56.45	33.38	19.26	1.19	79.55
2004	128.69	18.99	57.52	40.09	17.93	1.34	80.73
2005	160.30	21.54	56.74	54.39	26.88	1.39	71.73
2006	194.91	23.31	55.70	59.78	13.55	1.47	84.98
2007	232.70	20.60	59.44	77.15	10.16	1.42	88.42
2008	247.82	19.55	61.17	67.19	10.40	1.34	88.26
2009	214.24	20.01	59.25	45.13	11.36	1.80	86.84
2010	282.51	20.96	60.79	53.76	9.16	2.03	88.81
2011	293.69	20.43	62.04	57.03	10.48	1.78	87.74
2012	290.00	21.26	59.22	48.77	9.09	1.87	89.04
2013	278.77	22.42	58.23	49.12	10.73	1.93	87.34

Source: Eurostat (2016a, b). Exports: in millions of tonnes

As far as exports, China's modal split is turning slowly towards carbon intensive modes of transport (Table 5.4); air and ocean modes are key for exporting products, the former mode is key in monetary value and less so in physical units. Table 5.4 describes changes in export values (billion euros in 2014) in 1999–2013 for the China–EU trade relationship.

For example, in value terms the sea mode takes an important position by 2013 while air trails behind. In weight terms, shipping takes the lion share of exports more decisively than the air mode does. Today most of the exports out of China are on average of a lower value than those of the EU or the USA which reduces demand for air transport. In the future, however, if high-value products continue to be traded into or out of China, it is likely that airfreight flows will continue its upward trend which will equal levels of China's trading partners. Using the data in Table 5.4 we can detect changes in the weight value ratio in 2000–2013. A fall in the weight value ratio indicates more competitiveness for air cargo for two reasons (Hummels 2007). First, the marginal fuel cost of air cargo is higher than that of sea cargo. Second, consumers are sensitive to changes in the advertised price and not to those of transport cost. The improving competitiveness of air transport is reflected in the falling share of the sea mode (Table 5.4).

In contrast to China–EU trade, US–EU trade relies far more on the airfreight mode because of two reasons. First, the USA holds an advantage in high-technology industries, and in those exports. Second, the high share of air cargo in the US–EU trade reflects the fact that the EU region trades mainly with itself (including high-technology goods) and less with the US and non-EU regions. Table 5.5 depicts exports by the air and sea modes where clearly air transport leads as the key mode in value terms.

Table 5.5 US–EU export by transport mode

Year	Export (billions of euros)	By air (%)	By sea (%)	Export (millions of tonnes)	By air (%) (weight)	By sea (%) (weight)
1999	165.86	49.82	27.91	71.07	1.14	85.83
2000	206.28	52.90	25.19	72.45	1.12	85.03
2001	203.30	51.72	26.02	68.17	1.22	86.07
2002	182.62	48.59	27.94	62.08	1.13	87.21
2003	158.12	49.32	28.42	60.37	1.01	85.89
2004	159.37	49.27	28.45	63.49	1.44	87.44
2005	158.85	49.43	28.90	60.67	1.17	88.03
2006	170.36	46.12	31.49	63.02	1.29	87.69
2007	177.06	46.15	32.95	68.86	1.08	89.32
2008	182.35	45.91	36.29	88.24	0.84	88.29
2009	154.86	47.68	32.30	69.21	0.80	88.42
2010	173.07	51.09	34.81	76.71	0.90	91.03
2011	191.56	48.56	37.75	91.14	0.80	90.07
2012	205.29	48.00	38.00	100.29	0.65	90.01
2013	196.13	45.71	38.98	103.14	0.60	91.12

Source: Eurostat (2016a, b)

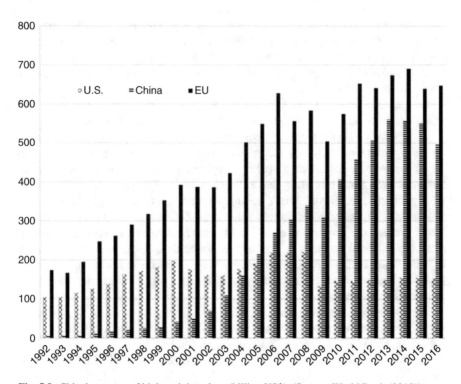

Fig. 5.9 China's exports of high-tech in values (billion US$). (Source: World Bank (2018))

The weight to value ratio of air cargo has fallen more in the US–EU trade link than that of China–US trade. This indicates that changes in trade patterns have favoured air transport in the US–EU trade route far more than those affecting air transport of the China–US route.

In Fig. 5.9 we can see that China and the EU export more high-tech products than does the USA; China has outperformed the latter in the review period. These exports include products with high R&D intensity, such as in aerospace, computers, pharmaceuticals, scientific instruments and electrical machinery.

In Fig. 5.10 we can see that China's infatuation with imports of high-tech (chemicals, basic manufactures, machinery and transport equipment, miscellaneous manufactured goods, luxury goods and gold) and imports of French cheese ought to bear an influence on airfreight movements. Another driver stemming from the import and export relationship is the fraction of manufacturing imports in total imports which is high for the three economies, but the USA shows a higher dependence of imported manufactures (in %) than China and the EU do (Fig. 5.10). These trends on imports inevitably bear on air transport demand between China and the USA and between the former and the EU.

Exports of manufactures comprise the commodities in SITC sections 5 (chemicals), 6 (basic manufactures), 7 (machinery and transport equipment), and 8 (miscellaneous manufactured goods), excluding division 68 (nonferrous metals).

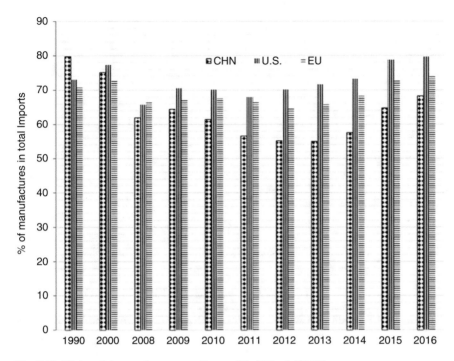

Fig. 5.10 High-tech imports by country. (Source: World Bank (2018))

5.4.3   Airfreight: China and the World

Figure 5.11 allows the reader to visualize changes of magnitude in airfreight volumes moved for the three economies. This graph is useful to understand the rapid changes in China's air cargo market.

In 1974–2012, the annual growth rate of China air cargo is much higher than that of the USA (4.5%) or Japan (6%). During 1974–1990 period, the average growth rate of China's air cargo is higher than that of the USA or Japan. Figure 5.12 shows per capita freight activity for the three countries. China has caught up quickly with the other two trading regions/partners.

By 2020 China is likely to match the US level of air cargo activity. Convergence in the growth rates of the three countries can be seen.

Figure 5.13 displays air cargo per capita for three countries in absolute levels before and after China's entry into the World Trade Organization.

On a per capita (absolute) basis, China's air cargo still lags behind that of the USA or Japan. In terms of growth rates, Japan shows rapid growth in the 1970s, slower growth in the 1980s and flat growth in the following decades (Fig. 5.12). China's air cargo starts from a lower base, and so its growth is faster than its competitors in most of the period. Air cargo volume levels off in recent years in the three countries because of falling activity of world trade after the 2008 financial crisis.

Putting together, Figs. 5.11, 5.12 and 5.13 shows that there is plenty of slack for the expansion of China's airfreight flows in the next decades: its airfreight level lies

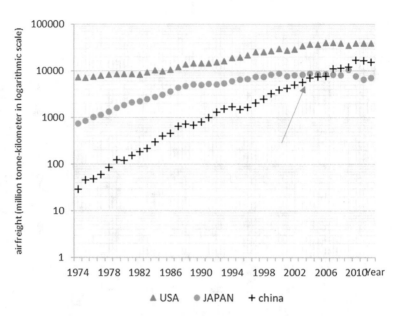

Fig. 5.11 Rate of growth of airfreight in selected economies. (Source: World Bank (2018))

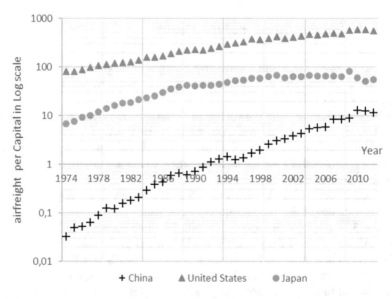

Fig. 5.12 Rate of growth of per capita airfreight in selected economies. (Source: World Bank (2018). In 2007 China's airfreight flows exceed 1975 Japan)

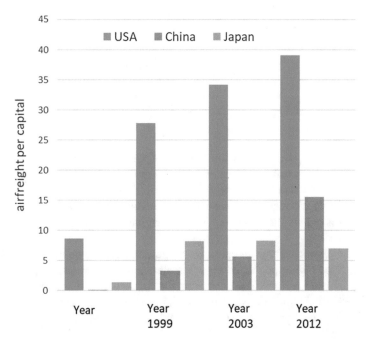

Fig. 5.13 Airfreight per capita before and after China's WTO entry. (Source: World Bank (2018))

below the USA and above Japan's. The paradox of the latter nation (Fig. 5.10) is that it has been a global trader but its airfreight volume per capita lies below that of China since 2012. Before the 2000s, China's low levels of airfreight activity can be explained because 1) it had not yet entered into the World Trade Organization 2) China was a closed economy and 3) did not trade with the outside world as much as it does today. A strong increase in airfreight activity after 2000 until today can be detected.

5.4.4   The Spatial Distribution of Air Cargo in China

In the above sections, we evaluated the broad relationship between international trade and both domestic and airfreight activities, we now examine the spatial features of airfreight activity at each of the key cities in China. The network characteristics of air cargo are examined elsewhere in the chapter.

The correlations for airfreight flows and for economic indicators are shown in Table 5.6 based on the sample for all of the cities. Section 5.6 describes further these correlations at the regional level. It is clear that the volume of airfreight is positively correlated to wages because the export dynamic raises wage levels or because high-tech exports are correlated to higher than average wages in cities. Exports of time-

Table 5.6 Airfreight
correlations and key
economic indicators

	Air cargo (in %)
Exports	17.4
Population density	32.8
Time sensitive product exports	40
Wages	54

sensitive products also explain the volume of airfreight and population density
appears as a less powerful explanation for airfreight growth. In theory, some level of
population density should be necessary to raise the demand and the production for
high-technology products, for time-sensitive products and other items that would
need to be moved via the air mode. Further correlation analysis is reported in Sect.
5.6, Table 5.19.

Table 5.6 shows average air cargo and wages, exports and population density.

As expected highly populated regions generate higher air cargo flows than less
densely populated regions: The Eastern region shows higher than average air cargo
flows, exports and population density. Population density reflects market size which
in turn influences goods consumption and goods production (manufacturing
activities).

5.5   The Primacy of Cities and Air Cargo

To explore the primacy of cities, from the perspective of air cargo flows, we separate
the concentration of air cargo from its main cities that generate those air cargo
flows.

5.5.1   Concentration of Airfreight by Regions

Tables 5.7 & 5.8 shows the concentration of air cargo flows by city hierarchy.

Shanghai concentrates 30% of total air cargo in Eastern China. The first top three
cities concentrate more than half of total air cargo in this region, the first five cities
concentrate almost three quarters and the top 10 cities almost all of the air cargo
flows (Table 5.8). The dominance of Shanghai points that urban interactions, of
airfreight, will be biased in a model of spatial interaction of air cargo.

Half of the cities in our sample, mostly small cities, represent only 1.3% of total
air cargo flows. In short, air cargo is unbalanced in its urban setting. This means air
cargo is not well defined by urban size or city size.

Table 5.7 Average air cargo: outbound (tonnes per year) Year: 2014

Region in China	Average air cargo (in 1000 tonnes) per city	Average exports (289 cities) 100 million	Average population density, people per sq km (289 cities)
Eastern	226.7	348.3	708,83
Western	42.2	29.62	*210.89*
Central	25.9	31.29	364.75

Year 2014

Table 5.8 Concentration of air cargo by selected cities in the Eastern region

	Air cargo (1000 tones)	% total in Eastern region	Average population density (people per sq km)
Shanghai	2938.16	30.8	2251
Top 3 cities	5986.78	62.8	1382
Top 5 cities	7271.58	76.3	1567
Top 10 cities	8495.62	89.1	1197

We can say the high concentration of airfreight flows in a few cities as described above reflects a possible law of Zipf (1949); that is they follow a power law behaviour, or as described in Cristelli (2012) 'a rank-size rule capturing the relation between the frequency of a set of objects or events and their size'. But further analysis is needed to establish to what degree the Zipf's law is applicable. The underlying mechanism of that law however needs to be defined. The data on airfreight and on exports are uneven: as a few cities concentrate exports the same cities will tend to concentrate airfreight flows as we have said above.

Tables 5.9 to 5.14 describe the relationship between airfreight movements, freight rank, exports and its rank by type of city and by region (Eastern, Central and Western China). A straight line between airfreight flows and exports can be detected in the data (Table 5.9) for most of the cities in our sample. In the Eastern region, the top exporter is Beijing city and Sanya the worst one. In the Central region, the top exporter is Zhengzhou city and Zhangjiaje the worst one. In the western region, Chengdu is the top city and Zhaotong the worst one.

We can detect however that a few cities manage to break the rule of a positive relationship between exports–airfreight activity. Some cities exhibit a steeper slope of exports but produce relatively smaller airfreight levels, perhaps because these cities are less technologically advanced and depend more on primary activities i.e. agriculture. Other cities show higher than average airfreight volumes along with lower levels of exports: trade volume does not always stimulate airfreight activity. For instance, the city of Shenyang ranks 29th in the export dimension but achieves the 12th rank in airfreight movement and so trade matters less in the case of that city.

Table 5.9 Airfreight and exports of Eastern China

City	Airfreight (1000 tonnes)	Freight rank	Export (100 million US$ 2015)	Export rank	Population density
Shanghai/Pudong	2938.16	1	2067.3	2	2251
Beijing/Capital	1799.86	2	596.32	5	791
Guangzhou	1248.76	3	589.15	7	1106
Shenzhen	854.9	4	2713.56	1	1440
Shanghai/ Hongqiao	429.81	5	2067.3	2	2251
Hangzhou	338.37	6	412.62	11	423
Xiamen	271.47	7	454	9	1214
Nanjing	248.07	8	319.01	14	969
Tianjin	194.24	9	483.13	8	845
Qingdao	171.89	10	407.91	12	682
Dalian	136.55	11	346.82	13	469
Shenyang	131.93	12	59.65	29	558
Haikou	99.94	13	17.98	40	701
Fuzhou	96.95	14	211.3	17	502
Wuxi	84.03	15	413.13	10	1016
Jinan	74.07	16	57.14	30	745
Ningbo	61.66	17	614.45	4	589
Sanya	52.6	18	0.52	42	298
Wenzhou	49.71	19	176.96	20	674
Shijiazhuang	39.66	20	73.38	27	634
Yantai	37.23	21	283.59	15	473
Quanzhou	35.71	22	123.75	22	629
Beijing/Nanyuan	30.05	23	596.32	5	791
Weifang	17.08	24	109.68	23	544
Zhuhai	16.27	25	216.37	16	618
Nantong	11.1	26	187.86	19	956

Source: China City Statistical Yearbook

We examine the stability of the data (Tables 5.9 to 5.14) by using standard deviation values of our sample. The standard deviation of airfreight flows in the two regions lies above its average as a result of volatility in the sector via the export sector. The high volatility makes it difficult for analysts to forecast the volume of exports and thereby of airfreight flows. Standard deviation values for export volumes are highly volatile in all regions. Standard deviation values for population density reveal that the density is less volatile than the other indicators.

As for Central China (Tables 5.11 and 5.12), Zhengzhou concentrates almost one fourth of that region's total air cargo flows: air cargo flows are more dispersed in this region than in the Eastern region. The first top three cities concentrate one third of the total, and the five cities concentrate over a third while the top ten cities take

Table 5.10 Airfreight movement and exports: Eastern China (continued from Table 5.9)

City	Airfreight (1000 tonnes)	Freight rank	Export (100 million US$ 2015)	Export rank	Population density
Changzhou	11.06	27	199.6	18	834
Jieyang	10.65	28	38.1	33	1280
Xuzhou	6.07	29	62.88	28	880
Weihai	5.09	30	106.59	24	437
Taizhou	4.38	31	172.39	21	628
Linyi	4.1	32	38.97	32	630
Yancheng	2.84	33	34.65	35	485
Yiwu	2.7	34	90.05	26	682
Zhanjiang	2.43	35	22.09	38	589
Dandong	1.4	36	28.75	36	157
Lianyungang	1.39	37	36.01	34	671
Jinzhou	1.12	38	17.6	41	311
Qinhuangdao	0.61	39	24.88	37	373
Dongying	0.48	40	49.82	31	233
Quzhou	0.47	41	18.59	39	286
Zhoushan	0.43	42	92.24	25	668
Average	**226.7**		**348.39**	**N.A**	**745.5**
Standard deviation	**554**		**580.3**		**439.0**

Table 5.11 Concentration of air cargo by selected cities in the Central region

	Air cargo	% total in Central region	Average population density
Zhengzhou	151.9	25.99%	1440
Top 3 cities	274.71	47.01%	988.6
Top 5 cities	426.87	73.05%	704
Top 10 cities	559.55	95.75%	588.6

almost half of the total. The rest of the cities concentrate less than 2% of the total. There are 18 small cities out of 28 cities in the region. In sum out of the three regions, Central China displays the highest dispersion of air cargo activity (Table 5.11).

One explanation for this is that the spatial distribution of manufacturing, high technology and other products may be even higher than in other regions. The *East* and *West* regions of the country reflect an unbalanced distribution of economic activity, trade and because of this of air cargo flows.

In Western China the city of Chengdu takes one third of total air cargo flows (Tables 5.13 and 5.14). The top three cities take two thirds of the total, the top five almost three fourths and the top ten almost all of the air cargo flows. This reflects a pattern of spatially clustered air cargo flows. The next group of smallest cities take

Table 5.12 Airfreight and exports of Central China

City	Airfreight (1000 tonnes)	Freight rank	Export (100 million US$ 2015)	Export rank	Population density
Zhengzhou	151.19	1	202.9	1	1440
Wuhan	128.2	2	107.48	3	967
Changsha	110.61	3	51.74	5	559
Harbin	85.95	4	15.51	12	187
Changchun	66.21	5	29.11	8	367
Hefei	42.6	6	136.28	2	621
Taiyuan	42.26	7	42.42	7	524
Nanchang	37.86	8	64.66	4	686
Yanji	5.39	9	1.88	27	300
Ganzhou	4.57	10	28.43	9	235
Yichang	4	11	16.63	11	189
Yuncheng	2.43	12	3.88	22	366
Zhangjiajie	2.22	13	0.36	28	179
Datong	2.13	14	2.34	26	226
Huangshan	1.88	15	6.27	21	150
Manzhouli	1.71	16	2.5	25	232
Changzhi	1.62	17	8.76	18	242
Mudanjiang	1.47	18	49.22	6	68
Enshi	1.42	19	3	24	168
Luoyang	1.06	20	11.89	15	466
Xiangyang	1.03	21	10.65	16	301
Jingdezhen	1.02	22	12.13	14	317
Qiqihar	0.86	23	6.35	20	132
Nanyang	0.81	24	8.18	19	455
Jiamusi	0.62	25	15.16	13	73
Anqing	0.25	26	10.05	17	405
Heihe	0.17	27	25.23	10	25
Changde	0.16	28	3.21	23	33
Average	**24.9**		**31.29**		**354**
Standard deviation	**43.4**		**46.6**		**303**

Source: China City Statistical Yearbook

about 9% of the total. These small cities are about 30 cities, out of 41 cities. The data show that the urban network of air cargo is also highly unbalanced within this region.

Inspecting the data for the Eastern region, we find that on average export activity of 348 (100 million US$, 2015) and an average population density level of 745 (person per kilometre) will explain the airfreight flows of 226.7 (1000s tonnes) in 2014. In the Central region an average expoert value (of 31.29) and an average population density (of 354) will generate relatively lower levels of airfreight flows

Table 5.13 Concentration of air cargo by selected cities in the western region

	Air cargo (1000 tonnes)	% total in Western Region	Average population density (people per sq km)
Chengdu	508.03	29.36	968
Top 3 cities	1038.94	60.04	544
Top 5 cities	1344.99	77.73	521
Top 10 cities	1601.09	92.53	382.8
1/3 smallest cities			150.5

Source: China City Statistical Yearbook (2015)

(1000s tonnes). In the Western region comparatively lower export volumes and a lower population density will translate into high volumes of airfreight (Table 5.14).

5.5.2   Air Cargo and Population Density

The data on population density (Tabulated in Tables 5.9 to 5.14) is highly uneven. The highest density levels are found in the Eastern region and the lowest in the Western one. In terms of variability of the region, we can see the latter region shows more variability between high population density and a lower one. Both the central and western regions show high variability. The high variability reflects both an unbalanced economic development and an uneven manufacturing activity as well as differing values of urban land. (High population density is a proxy of high land values.) At the province level these reflect city size which varies from the smallest province of 3.1 million (Tibet) to 107.2 million (Guangdong province).

As elsewhere, population density in China's cities is a key influencer in the growth of airports because they represent catchment areas of the economic activity of airports. In order to visualize to what extent airports serve the population within China's counties and within the regions considered (Tables 5.9 and 5.14), economists estimate the proximity of populations to the nearest airport. Wang and Jin (2007) use counties as the basic units of statistical analysis to identify parts of the country served by airports. Two ranges of distance are used, 50 and 100 km to identify the degree to which counties are served. Wang assigned a city to a hinterland's nearest airport facility.

Table 5.15 shows the serviced areas, population served by airfreight services and its related GDP. The table shows that the Eastern area offers the better coverage of airfreight services to its population (39%) in contrast to the other regions, for example, the Central region only serves 24% of its population. The Eastern area offers superior coverage than the rest of the mainland China.

To summarize Tables 5.9 to 5.15 the following is found. Although further information among airfreight flows (flows among city pairs) is needed in our sample, the

Table 5.14 Airfreight and exports of Western China

City	Airfreight (1000 tonnes)	Freight rank	Export (100 million US$ 2015)	Export rank	Population density
Chengdu	508.03	1	303.64	2	968
Chongqing	268.64	2	385.68	1	406
Kunming	262.27	3	56.86	5	259
Xi'an	174.78	4	72.99	4	787
Urumqi	131.37	5	80.64	3	187
Guiyang	79.59	6	42.14	6	466
Nanning	78.13	7	25.17	10	321
Lanzhou	35.95	8	26.92	9	246
Guilin	33.76	9	7.89	21	188
Hohhot	28.67	10	8.33	20	132
Yinchuan	26.9	11	10.98	15	185
Lhasa	15.34	12	32.61	8	17
Xining	15.28	13	6.62	22	259
Baotou	9.27	14	11.64	14	80
Lijiang	6.95	15	0.85	31	56
Liuzhou	6.21	16	9.07	18	200
Mianyang	4.94	17	13.73	11	269
Xishuangbanna	4.89	18	9.65	17	60
Dehong	4.88	19	12.76	12	107
Kashi	4.74	20	10.74	16	37
North Sea	4.26	21	11.84	13	504
Hailar	4.05	22	2.77	26	11
Yibin	2.76	23	5.42	23	412
Yulin	2.41	24	0.41	36	86
Korla	2.35	25	0.30	38	75
Luzhou	2.14	26	1.58	28	413
Yining	1.67	27	37.59	7	52
Tongliao	1.49	28	0.73	34	54
Nanchong	1.38	29	3.98	25	609
Chifeng	1.24	30	1.27	30	51
Wuhai	1.22	31	0.02	41	313
Aksu	1.11	32	5.36	24	19
Jiayuguan	0.95	33	0.34	37	67
Pu'er	0.69	34	1.41	29	55
Dali	0.6	35	1.66	27	121
Lincang	0.5	36	0.75	33	97
Diqing	0.46	37	0.10	39	17
Baoshan	0.37	38	0.78	32	130
Altay	0.13	39	9.06	19	6
Yan'an	0.06	40	0.66	35	64
Zhaotong	0.04	41	0.02	40	258

(continued)

Table 5.14 (continued)

City	Airfreight (1000 tonnes)	Freight rank	Export (100 million US$ 2015)	Export rank	Population density
Average	**388**		**29.62**	**40**	**210.8**
Standard deviation	**169.4**		**75.6**		**216.3**

Source: China City Statistical Yearbook (2015)

Table 5.15 Area, population, GDP according to rational service hinterlands, in percentages

(In %)	50 km			100 km		
Indicator	Area	Population	GDP	Area	Population	GDP
Eastern	27.7	39.2	63.8	66	69.3	85.9
Central	14.4	24.2	31.7	44.6	56.5	56
Western	13.2	27.4	46.3	28.7	55.7	68.8
Mainland China	15	31.1	52.6	37.2	61.5	75.4

Source: Wang and Jin (2007). Year: 2002

Table 5.16 Concentration of links by selected cities in the Eastern region

	Air cargo routes	% total in Eastern region
Eastern region		
Shanghai	340	19
Top 3 cities	600	34
Top 5 cities	825	47
Top 10 cities	1089	62

Source: China City Statistical Yearbook (2015). Elaborated by the author

data show that the network of key cities concentrates most of the air cargo moved. Our data do not allow us to observe the changes in the ATNC in terms of the strength of links among cities, nodes or edges but the data does lead us to conclude that there is large variation among cities.

5.6 Airfreight Flows and Connectivity of Cities

Having defined the spatial concentration of air cargo among key Chinese cities in the section above, we aim to (1) find whether the high connectivity of a network influences airfreight volumes in the ATNC, (2) unravel if the links among spatial networks and airfreight activity explain how vital the sector is and (3) explore the hierarchy of cities as far as connectivity. This analysis relies strongly on the data described in the previous tables (Tables 5.9 through 5.16).

The data on connectivity of cities confirms that the pattern that emerges is the following: the greater the connectivity the greater the air cargo volume that results from trade among cities and regions (Tables 5.20 to 5.23.)

Table 5.17 Concentration of links by selected cities in the Central region

	Air cargo routes	% total in Central region
Central region		
Wuhan	179	16
Top 3 cities	505	46
Top 5 cities	67	61
Top 10 cities	930	84

China statistics, elaborated by the author. China City Statistical Yearbook (2015)

Table 5.18 *Concentration* of links by selected cities in the Western region

	Air cargo routes	% total in Western region
Western region		
Chengdu	85	10.83
Top 3 cities	230	29.30
Top 5 cities	373	47.50
Top 10 cities	530	67.52

China City Statistical Yearbook (2015). Elaborated by the author

Table 5.19 Correlation matrix (%) among airfreight transport, exports and population

Region	Exports by city (%)	Population density by city (%)	Connectivity per region (# of routes) (%)
(Eastern region) Airfreight flows	0.64	0.61	0.825
(Central region) Airfreight flows	0.76	0.78	0.78
(Western region) Airfreight	0.13	0.29	0.21

Density and connectivity based on data for the whole sample of China's cities. Source: China City Statistical Yearbook (2015). Elaborated by the author

It is important for the purposes of this chapter to identify China's equivalent to the city of Memphis in the USA. Because of its aerial geography that city has taken a leading place in the US air cargo market. It has been said that Memphis was chosen because 'the hub has to sit somewhere in a trapezoid between Memphis in the Southwest, to Champaign Illinois, in the Northwest, over to Dayton, Ohio, and down to Chattanooga. It has to sit in that footprint' (Kasarda and Lindsay 2011, p. 61). Memphis is able to serve every key node or city in the USA overnight, and it has an equidistant position to every major US city. It lies inside an advantageous time zone in the USA. One possible candidate is the Chinese city of Wuhan because 'China's

Table 5.20   Airfreight connectivity and population density: Eastern China

City	# of routes	Routes rank	Population density	Population density rank
Shanghai/Pudong	170	2	2251	1
Beijing/Capital	260	1	791	13
Guangzhou	147	4	1106	6
Shenzhen	78	5	1440	3
Shanghai/Hongqiao	170	2	2251	1
Hangzhou	60	7	423	36
Xiamen	52	10	1214	5
Nanjing	55	9	969	8
Tianjin	51	11	845	11
Qingdao	46	12	682	17
Dalian	73	6	469	34
Shenyang	59	8	558	29
Haikou	37	14	701	16
Fuzhou	29	19	502	31
Wuxi	18	22	1016	7
Jinan	37	14	745	15
Ningbo	25	20	589	28
Sanya	34	16	298	39
Wenzhou	30	17	674	19
Shijiazhuang	30	17	634	22
Yantai	22	21	473	33
Quanzhou	14	24	629	24
Beijing/Nanyuan	39	13	791	13
Weifang	8	32	544	30
Zhuhai	14	24	618	26
Nantong	9	30	956	9
Average	**60.2**	**NA**	**852**	**NA**

China City Statistical Yearbook (2015)

equivalence to Memphis is the Wuhan urban region. The triple city of Wuhan has a geographical centrality that gives its site immense strategic and commercial significance. Located at the very heart of China, it is roughly equidistant from the cities of Beijing and Guangzhou (Canton) on a north south axis and also equidistant from shanghai and Chongqing on an East west line' (Pletcher 2011, p. 230).

Furthermore, Wuhan is crossed by 'converging maritime, river, rail and road transportation routes from almost every direction' (Pletcher 2011). The intermodal competition among rail, water and air has ensured that Wuhan has not become the centralized node of airfreight flows.

Tables 5.16 to 5.18 show the concentration of connectivity of airfreight movements within the regions. Airfreight activity and connectivity at the regional level reveal that in general the Eastern region is better connected than the other two regions as we have already found in the preceding section. The Eastern region shows strong bias for Shanghai as the number of links among airports increases. Three

Table 5.21 Airfreight and connectivity: Eastern China (continued)

City	# of routes	Routes rank	Population density	Population density rank
Changzhou	11	27	834	12
Jieyang	18	22	1280	4
Xuzhou	14	24	880	10
Weihai	7	34	437	35
Taizhou	8	32	628	25
Linyi	11	27	630	23
Yancheng	4	38	485	32
Yiwu	9	30	682	18
Zhanjiang	5	36	589	27
Dandong	2	41	157	42
Lianyungang	10	29	671	20
Jinzhou	5	36	311	38
Qinhuangdao	2	41	373	37
Dongying	4	38	233	41
Quzhou	3	40	286	40
Zhoushan	6	35	668	21
Average: Tables 5.20 and 5.21	**40**	**NA**	**745**	**NA**
Standard deviation	**54.3**		**444**	

cities account for most of the connectivity levels in the Eastern region; three cities account for half of connectivity at the Central region. The top three cities in the Western region account for one third of total connectivity. The dominance of the top three cities in the Eastern region is the lowest of all regions (Table 5.16). Similarly, the dominance of the top five cities is lowest in the Eastern region in comparison with the other regions.

The dominance of the top 10 cities in terms of their concentration of links varies from 62 to 84% (three regions). This means that the cities in the Eastern region are more evenly connected to each other. In other words, the Eastern region has on average more links connecting each Eastern city. Cities within the Central and Western regions are relatively less well connected and have developed fewer air arteries with their own hinterland (Tables 5.17 and 5.18).

Table 5.19 depicts the correlation matrix using the data at hand in Tables 5.20, 5.21, 5.22 and 5.23. Theory holds that the high connectivity produces economic growth which is confirmed in value of correlation level between airfreight flows and connectivity is high in two regions: both for the Eastern and Western ones (Table 5.19). The least dynamic region, the Western region, shows low correlation values in all dimensions.

Table 5.20 tabulates the connectivity by city (no. of routes) of airfreight lows, routes rank and population density. The least dynamic cities lie in the heartland (Central region, Table 5.22).

How much effect have spatial factors had on the airfreight flows?

Table 5.22 Airfreight connectivity: Central China

City	# of routes	Rank	Population density	Rank
Zhengzhou	149	3	1440	1
Wuhan	179	1	967	2
Changsha	177	2	559	5
Harbin	106	4	187	21
Changchun	65	7	367	10
Hefei	63	8	621	4
Taiyuan	94	5	524	6
Nanchang	76	6	686	3
Yanji	11	14	300	15
Ganzhou	10	18	235	17
Yichang	22	9	189	20
Yuncheng	20	10	366	11
Zhangjiajie	11	14	179	22
Datong	11	14	226	19
Huangshan	13	11	150	24
Manzhouli	8	22	232	18
Changzhi	12	12	242	16
Mudanjiang	10	18	68	27
Enshi	4	26	168	23
Luoyang	10	18	466	7
Xiangyang	12	12	301	14
Jingdezhen	11	14	317	13
Qiqihar	6	24	132	25
Nanyang	7	23	455	8
Jiamusi	9	21	73	26
Anqing	3	27	405	9
Heihe	3	27	25	28
Changde	5	25	333	12
Average	**39**	**NA**	**364**	**NA**
Standard deviation	**53.8**	**NA**	**296**	**NA**

China City Statistical Yearbook (2015).

Two spatial factors that stimulate freight flows include (1) connectivity among China's cities and (2) population density for the three regions: the Eastern, Western and Central ones. For the regions we find that airfreight activity and connectivity are highly correlated (using data sets of Tables 5.20, 5.21, 5.22 and 5.23).

The above finding leads us to ask: How much effect has air transport and freight activity had on the economy? Our data suggest that the sector has stimulated the economy as follows: First connectivity achieves higher exports via air transport and the latter, in turn, puts pressure on greater connectivity among cities. The two effects feed on each other.

The second effect on the economy is the effect of population density (already discussed in Sect. 5.5) which signals a) the size of the labour force and b) the market for

Table 5.23 Airfreight connectivity: Western China

City	# of routes	Routes rank	Population density	Population density rank
Chengdu	85	2	968	1
Chongqing	75	3	406	8
Kunming	88	1	259	13
Xi'an	67	4	787	2
Urumqi	58	5	187	18
Guiyang	37	6	466	5
Nanning	31	7	321	9
Lanzhou	29	9	246	15
Guilin	29	9	188	17
Hohhot	31	7	132	20
Yinchuan	22	11	185	19
Lhasa	12	17	17	38
Xining	19	12	259	12
Baotou	15	13	80	26
Lijiang	13	16	56	31
Liuzhou	14	14	200	16
Mianyang	14	14	269	11
Xishuangbanna	6	21	60	30
Dehong	489	18	107	
Kashi	3	30	37	23
North Sea	3	30	504	36
Hailar	12	17	11	4
Yibin	10	19	412	40
Yulin	6	21	86	7
Korla	6	21	75	25
Luzhou	2	33	413	27
Yining	6	21	52	6
Tongliao	2	33	54	34
Nanchong	8	20	609	33
Chifeng	4	28	51	3
Wuhai	5	25	313	35
Aksu	3	30	19	10
Jiayuguan	1	36	67	37
Pu'er	1	36	55	28
Dali	1	36	121	32
Lincang	4	28	97	22
Diqing	1	36	17	24
Baoshan	5	25	130	39
Altay	5	25	6	21
Yan'an	1	36	64	41
Zhaotong	2	33	258	29
Average	19	NA	210	NA

China City Statistical Yearbook (2015).

airfreight products and its services. The relationships "a" and "b" versus density are confirmed empirically: density & airfreight activity are positively correlated in our sample in the three regions (Tables 5.19 to 5.23). This is because density is affected by the the catchment area for each airport; the area, in turn, boosts airfreight movement.

A further effect on the economy of population density is its effect on land values and on high manufacturing employment both of which are key features of high technology exports and thus of airfreight flows. Therefore the data shows that density is a useful predictor of airfreight flows.

Using the data sets at hand, we calculate the averages for connectivity and population density of the cities in each region (Tables 5.20 to 5.23). The top four cities show higher than average indicators.

As mentioned above as far as connectivity, the Eastern region shows the highest levels of connectivity and the standard deviation values are low, reflecting a low volatility for connectivity. The lowest level of connectivity can be seen in the West of China, but that region shows a higher volatility in this dimension than the other regions.

There are 42 cities in the Eastern region averaging 745 people per square kilometre. In the Western region there are 41 cities averaging 210 person per square kilometre and in the central one 28 cities averaging 364. These are significant differences in population density among the regions which will impact on the shape of the ATNC network. The data sets show that density varies significantly from region to region and from city to city. It is feasible to argue that there should be an optimum level for population density to sustain either a larger or lower volume of airfreight flows which will depend on market size (city size).

As the pace of urbanization increases in China, it is feasible to expect higher population density in many cities which will increase the demand for both airfreight services and greater connectivity of the ATNC.

As said above, the ATNC consists of 285 cities and associated airports making up a network (of air cargo activity) that is highly polarized regionally: The network reflects an economic development patttern that is unbalannced. While Shanghai represents the most important air cargo interaction in the country, distorting air cargo in the ATNC, secondary interactions are established by other cities.

In the context of the concentration of links of the ATNC experts have described that "*the ATNC cumulative degree distribution follows an exponential function as p(k) 0 0.703e-0.047k that is to say, a few busy cities at the top dominate the system with a large number of routes and the number of routes to each city decline rapidly and level off towards small cities most of which have only 1–3 routes*" (Wang and Jin 2007).

The above statement can be taken to imply that the number of links of airfreight flows originating in the top 20 cities is concentrated. The Delta river of the Yangtze, the Zhujiang river Delta and the Jing Ji Jin regions generate most of the exports that require airfreight services (Table 5.24). Independent evidence shown in Tables 5.9 to 5.14 also confirm this level of dominance for all key cities for each region. These three regions generate a large chunk of GDP and exports and concentrate most of China's manufacturing activity.

Table 5.24 National share of area, population and GDP of the three nodes of China's triangular airfreight axis. year: 2007

Economic region	Airfreight volume (unit:1000s tones)	Share
Jing ji	1994.1	16.67
Yangtze river delta	3705.33	30.98
Zhujiang river delta	2103	17.58
Sum of three nodes	7802.43	65.25
Other regions	4154.49	34.74
Mainland China	11,956.92	100

Jing ji Ji: Beijing, Tianjin, Hebei province
The Yangtze river delta includes: Shanghai, Nanjin, Ningbo, Hangzhou, Huzhou, Wuxi, Suzhou, Zhenjiang, Tangzhou, Taizhou, Changzhou, Nantong, Jiaxing, Shaxing, Zhoushan The Zhujian river delta includes: Guangzhou, Senzhen, Zhuhai, Foshan, Jiangnen, Dongguan, Zhongzhan, Huzhou, Zhaoqinq
Source: Wang and Jin (2007)

5.7   Conclusion

Like cities do, airports are advancing on each other alongside air power and freight. This chapter examines airfreight transport networks from the optic of international trade and the spatial structure among China's cities. In the chapter, the ATNC network is superimposed on the urban network of China's cities bringing a new method to observe airfreight transport. The ATNC (a) supports export growth of high value products and (b) untaps the geographical advantages of manufacturing establishments within China and (c) strengthens the existing system of cities in China.

To examine what determines the period of 1975–2015, we closely examine key factors such as the type of trade, air transport deregulation, population growth and density, urbanization rates, proximity to markets and competition from the perspective of the air and rail modes and e-commerce. Further factors as far as the management of operations include demand for timely and speedy delivery of goods.

Our evidence suggests the historical growth of airfreight flows of China has tracked closely that of domestic and international trade. Our results suggest that the weight to value ratio has fallen significantly for the air mode which improves its competitiveness in the 1990–2015 period significantly.

Our results also suggest that the growth of trade (exports) stimulates airfreight flows. A key driver of air cargo trade has grown enormously in the last 15 years taking China from small player in international trade to a major exporter supported by the air mode. The share of air cargo in the China–US trade nexus is now much larger than in 2000 (in value terms), and the high-tech export sector has benefited from the advantage of air cargo although the shipping mode continues to be the main mode to export products from China in physical terms. In contrast to the China-US nexus, the China–EU trade nexus is less dominated by the air mode in value terms. The trade gains that the air mode provides are therefore extremely large. In short, airfreight activity can be explained by the import and export activity as well as by the geographical spread of manufacturing establishments which surround urban regions.

Our economic geography analysis of airfreight flows suggests that connectivity, population density and the law of Zipf explain the expansion of the sector.

We use the city as the unit of analysis of airfreight flows to examine the sector on three levels: (a) the spatial distribution of airfreight demand in China and its relationship to population density, (b) on how the network of cities in China affects airfreight through the effect of connectivity and (c) on how airfreight flows contribute to urban development and industrial growth. Six outcomes arise out of the analysis of air transport among cities. First, the results suggest that airfreight flows and population density are positively and highly correlated in contrast to surface freight transport. Further research is needed to explain the links among the density of cities and airfreight markets.

Second, we investigated the relationships among China's regions and air cargo traffic using a cross-sectional data set on airport cargo traffic, wages, population density, regional trade, high technology and time-dependent -exports. The socioeconomic variables exhibit a positive impact on air cargo in cities throughout China. We find that an increase in wages raises air cargo volumes significantly. Our results hold that a proportionate relationship between city size and air cargo holds as in many other studies. Average wages exhibit a positive relationship with air cargo.

Third, airfreight demand and flows can also be explained by the connectivity levels for a great number of cities, but the positive relationship does not always hold for all cities. The system of cities relies on the ATNC in recent years.

Fourth, airfreight demand can be explained by the population density since the latter reflects land values, market size and the labour force size.

Fifth, the rising tide of urbanization never seen before in modern history will likely increase the demand for air cargo, and simultaneously, the air cargo mode will create new urban regions by expanding the high-technology industry and opening up new markets.

Sixth, the pattern of airfreight demand from the viewpoint of major cities shows that the law of Zipf is observable, that is a few cities produce the largest part the airfreight demand. Although further analysis is required, we can observe that a few cities concentrate most of the airfreight flows and most of the connectivity capabilities which contributes to unbalanced economic development of China.

The evidence of 285 cities reveals that China's air transport network is heavily unbalanced, reflecting the geographical distribution of manufacturing, of high technology and of food production activities. These activities are mostly located in the East of the country. Although the airfreight sector has contributed to the emergence of sectors that depend heavily on airfreight transport, it is important to understand the geographical position of production and of high tech plants to predict the future geographical distribution of freight in order to better plan emerging networks and improve the efficiency of current ones, to better exploit (1) scale economies, (2) agglomeration economies and (3) network economies.

Two things emerge from our analysis of spatial trends. First, the distribution of air cargo is heavily clustered in cities of the east of the country. And, second the levels of connectivity among cities are uneven: the East of the country is the best connected region, making it a competitive region. One of the rationales for airport expansion is that this raises (a) the connectivity of trading relationships within and among regions and (b) the competitiveness of cities and regions. Our results suggest these two outcomes.

As China continues to expand its ATNC, its cities will continue to have impact on the worldwide air transport network and within China. It is important to identify which cities are poorly connected to the ATNC and this chapter has tried to do so. In the chapter, we have shown that there is a strong link between trade and urban growth confirming partially the Krugman and Livas (1996) argument.

Future research should focus on how the current trend for higher volumes of air cargo is likely to widen jet fuel demand. China is a net importer of oil, and so it needs a freight strategy that relies on low carbon transport systems such as the railways. However, China's increasing reliance on the air mode is raising the demand for oil imports, and at the same time, this mode is turning China into a mega exporter of high technology goods with substantial financial gains.

References

Acatitla E, Alonso J (2017) The use of complex networks in economics: scope and perspectives. Interdisciplina 5(12):9–22. (In Spanish)

Aschauer D (1990) Why is infrastructure important? Federal Reserve Bank of Boston. Conference Proceedings. 34: 21–68. http://www.bostonfed.org/economic/conf/conf34/conf34b.pdf. Accessed in Sept. 2014

Auden WH (1936) Poetry. https://www.theguardian.com/books/2007/dec/01/featuresreviews.guardianreview10

Banister D (2010) The trilogy of distance, speed and time. J Transp Geogr 19(4):950–999

Banister D, Berechman J (2000) Transport investment and economic development. Psychology Press, Hove

Banister D, Anderton K, Bonilla D, Givoni M, Schwanen T (2011) Transportation and the environment. Annu Rev Environ Resour 36:247–270

Barabasi L, Albert R (1999) Emergence of scaling in random networks. Science 286:509–512

Bookbinder J (2013) Handbook of global logistics. Transportation in international supply chains. Springer, New York, pp 1–547

Borges JL (1944) Ficciones. Everyman Press, London, pp 1–148

Cairncross F (1997) The death of distance. Harvard Business School Press, Cambridge

Camagni P (1993) From city hierarchy to city network: reflections about an emerging paradigm. In: Structure and change in society in the space economy. Springer, Berlin, pp 66–87

Castells M (1996) The rise of the network society. Vol 1 of the information age: economy, society and culture. Blackwell, Oxford

Castells M (2007) Communication, power and counter power in the network society. Int J Commun 1:238–266

China City Statistical Yearbook (2014/2015) Edited by China Statistical Press, pp 1–419

Choi JH, Barnett GA, Chon BS (2006) Comparing world city networks: a networks analysis of internet backbone and air transport intercity linkages. Global Netw 6(1):81–99

Clarke G, Martin R, Tyler P (2016) Divergent cities? Unequal urban growth and development. Cambridge J Reg Econ Soc 9(2):259–268

Congressional Research Service (2018) China-US trade issues. Report prepared by R. W Morrison, January 23, 2018, pp 1–71. 7-5700, RL33536, Washington, U.S.

CPC (Communist Party of China) (2016) The Thirteen five year plan for economic and Social Development of the People's Republic of China. Recommendations of the Central Committee of the Communist Party of China (CPC) for the 13th Five-Year Plan for Economic and Social Development of the People's Republic of China (2016–2020)

Cristelli M, Batty M, Petrionero L (2012) There is more than a power law in Zipf. Sci Rep 2:812

Dewey J (1927/1954) The public and its problems. Swallow Press. Athens. pp 1–236

Ducruet C (2017) Transport networks. International Encyclopedia of Geography: people, the earth, environment and technology, pp.

Erdos P, Renyi A (1960) On the evolution of random graphs.Publication of the Mathematical Institute of the Hungarian Academy of Sciences, 5:17–61. http://snap.stanford.edu/class/cs224w-readings/erdos60random.pdf. Accessed Feb 2017

Eurostat (2009) China passes the EU in High tech exports. Statistics in Focus

Eurostat (2016a) China-EU trade. By transport mode. Various years

Eurostat (2016b) U.S.-EU export by transport mode. Various years

Harvey D (1996) Justice, nature, & the geography of difference. Blackwell, Oxford

Hickman R, Givoni M, Bonilla D, Banister D (2015) Handbook of transport and development. Edward Elgar, Cheltenham

Hildeggun Kyvin N (2007) Time as a trade barrier: implications for developing countries. OECD Econ Stud 2006(1):137–167

Hummels D (2007) Transportation costs and international trade in the second era of globalization. J Econ Perspect 21(3):131–154

Hutchins D (1999) Just in time, 2nd edn. Gower, Aldershot

IBRD-World Bank (2009) World development report 2000: reshaping economic geography. World Bank, Washington, DC

IPCC (1999) Aviation and the global atmosphere. A special report of working groups I and III of the intergovernmental panel on climate change. Lead Authors: J.H. Ellis, N.R.P. Harris, D.H. Lister, J.E. Penner. http://www.ipcc.ch/ipccreports/sres/aviation/index.php?idp=2

Isard W (1945) The economic implications of aircraft. Q J Econ. 59(2):145–169

Jin F, Wang C, Xiuwei L, Wang J (2010) China's regional transport dominance: density, proximity and accessibility. J Geogr Sci 20:295–309

Johnson S (2014) How we got to now: six innovations that made the modern world. River Head Books/Penguin Group, New York, pp 1–289

Kansky K (1963) The structure of transportation networks: relationships between network geography and regional characteristics. Chicago: University of Chicago, 1963. Research paper no. 84

Kasarda J, Lindsay G (2011) Aerotropolis: the way we'll live next. Allen Lane/Penguin Books, London/New York, pp 1–449

Katz M, Shapiro C (1994) Systems competition and network effects. J Econ Perspect 8(8):93–115

Krugman P (1998) What's new about the new the economic geography? Oxford Rev Econ Policy 14(2):7–17

Krugman P, Livas R (1996) Trade policy and the third world metropolis. J Dev Econ 49:137–150

Lakew PA, Tok YCA (2015) Determinants of air cargo traffic in California. Transp Res Part A 80:134–150

Lao X, Zhang X, Shen T, Skitmore M (2016) Comparing China's City Transportation and Economic Networks. Cities 53:43–50

Liebowitz SJ, Margolis S (1995) Path dependence, lock-in, and history. J Law Econ Organ 11(1):205–226

Lin J (2012) Network analysis of China's Transportation Systems, statistical and spatial structure. J Transp Geogr 22:109–117

Mahutga MC, Ma X, Smith D, Timberlake M (2010) Economic globalization and the structure of the world city system: the case of airline passenger data. Urban Studies 47(9):1925–1947

Marx K (1857) The Grundisse (originally written in German). Penguin Books in association with New Left Review, 1973. https://www.marxists.org/archive/marx/works/1857/grundrisse/index.htm

McLane B (2012) A new history of documentary film. Bloomsbury Academic, New York, pp 73–92

McLuhan M (1964) Understanding media: the extensions of man. McGraw Hill, New York

Nagurney A, Li D (2015) Competing on supply chain quality: a network economics perspective. Springer, Heidelberg, pp 1–381

National Bureau of Statistics China (2016a) China's exports of manufactured goods (US$). Sourced from http://www.stats.gov.cn/english/Statisticaldata/AnnualData/

National Bureau of Statistics China (2016b) China's air freight volume (bn. T-km). Sourced from http://www.stats.gov.cn/english/Statisticaldata/AnnualData/

National Bureau of Statistics China (2016c) "Freight traffic by mode (million tonnes)". Sourced from http://www.stats.gov.cn/english/Statisticaldata/AnnualData/

National Bureau of Statistics China (2016d) Air Freight traffic by city (million tonnes)

Natonal Bureau of Statistics (2017) Statistics on airports

Nordas H (2007) Time as a trade barrier: implications for developing countries. OECD Econ Stud 1:137–167

Parnreiter C (2018) Geografia Economica: Una introduccion contemporanea. UNAM, Mexico City, pp 1–520. Edited by faculty of Economics

Pereira RAO, Derudder B (2010) Determinants of dynamics in the world city network, 2000–2004 [J]. Urban Stud 47(9):1949–1967

Pletcher K (2011) Understanding China: the geography of china, sacred and historic places. Britannica Educational Publishing and Rosen Educational Services LLC, New York, pp 1–273

Reynolds-Feingham A (2013) Comparative analysis of air freight networks in regional markets around the globe. In: Bookbinder J (ed) Handbook of global logistics, International series in operations research & management science, vol 181. Springer, New York

Rietveld P (1989) Infrastructure and regional development: a survey of multiregional economic models. Ann Regional Sci 23:255–274

Scheuerman W (2014) Globalisation. The Stanford Encyclopedia of Philosophy(Summer Edition) Edward N. zalta (ed), https://plato.stanford.edu/entries/globalization/. Accessed Feb 2016

Schivelbusch W (1978) Railroad space and rail road time. New German Critique 14:31–40

S.E.P. (2002) The Stanford Encyclopedia of Philosophy(Summer Edition) Edward N. zalta (Ed), https://plato.stanford.edu/entries/globalization/

Stokes C (1968) Transportation and economic development in Latin America. Frederick A. Praeger, New York, pp 1–204

Straitstimes (2017) China's rail ambitions run at full speed. Straitstimes, Sept 19

Taaffe E, Morrill R, Gould P (1960) Transport expansion in underdeveloped countries: a comparative analysis. Geogr Rev 53(4):503–529

Taylor P (2001) Specification of the World City Network. Geogr Anal 33(2):181–194

Taylor P, Catalano GD, Walker (2002) Measurement of world city network. Urban Studies, 2367–2376

UK Department of Transport (2005) Transport, the wider economic benefits and impacts on GDP. Discussion paper, pp 1–67

US Census Bureau (2017) US-China trade by Air (2017) FT 920 U.S. Merchandise Trade: selected highlights, U.S. Census Bureau, Department of Commerce, Suitland, MD, 1990-2007. https://www.census.gov/foreign-trade/Press-Release/2017pr/12/ft920/index.html

U.S. Census Bureau (n.d.) various years U.S. Exports of merchandise. http://www.census.gov/foreign-trade/reference/products/catalog/expDVD.html

Virilio P (1977) Speed and politics. Originally published by Edition Galilee, Paris. Republished by Semiotext (e) 2007, Los Angeles, USA, pp 1–174

Walras L (1874) Elements of pure economics, or, the theory of social wealth. Routledge, pp 1–619. Edition: Reprint (1 of decembre 2003)

Wang J, Jin F (2007) China's air passenger transport: an analysis of recent trends. Euroasian Geogr Econ 48(4):469–480

Wang J, Mo H, Wang H, Jin F (2011) Exploring the network structure and nodal centrality of Chinas air transport network: a complex network approach. Transp Geogr 19:712–721

Wang J, Mo H, Wang F (2014) Evolution of air transport network of China: 1930-2012. J Transp Geogr 40:145–158

Wang J, Bonilla D, Banister D (2016) Air deregulation and its impact on airline competition 1994-2012. J Transp Geogr 50:12–23

William S (n.d.), Globalization. The Stanford Encyclopedia of Philosophy (Summer 2014 Edition), Edward N. Zalta (ed.), URL=<https://plato.stanford.edu/archives/sum2014/entries/globalization/>

Woolf V (1928) Orlando. A harvest book. Harcourt, New York, pp 1–331

World Bank (2015) World development indicators. http://databank.worldbank.org/data/reports.

World Bank (2018) World development indicators. http://databank.worldbank.org/data/reports. aspx?source=2&country=CHN#. Various years

World Bank (2019) World integrated trade solution (2019). https://wits.worldbank.org/CountryProfile/en/Country/CHN/Year/2017#section2

Zipf GK (1949) Human behavior and the principle of least effort. Addison-Wesley, Cambridge

Chapter 6
Conclusions

The underlying message of this book is that there is a need for government intervention on a global scale for three reasons. First, to free up capacity of roads investment in other transport modes is urgently needed in all the major economies but particularly in China, the EU, and Mexico; government intervention is the only means to achieve large-scale investment. Second, there is a need for international institutions to regulate the freight transport sector particularly air transport related traffic and its environmental impacts. Third, to adopt policy measures for sustainable freight transport, governments need to regulate the sector in the areas of 1) fuel economy of new trucks, 2) rail locomotives, 3) aircraft engines and 4) vessel engines. One ought to bear in mind that certain regions i.e. China or Mexico need to rely on an expanding transport sector to stimulate economic growth and trade. So the winners of global trade have recently been the Asian middle class and the global rich (Banister 2018). In *'Air power and freight: The view from the European Union and China'*, we build on insights, discoveries and theories of great economists, transport economists, geographers and urban planners to examine the recent past and to return to the future of freight transport.

One can speculate that the freight market is dematerializing and undergoing a transformation by offshoring of manufacturing capacity. This industrial flight correlates with lower road freight and air transport activity in the coming decades.

These series of essays are the only studies that examine the modern history of road freight transport and air transport in combination with recent decades for several fast-growing (i.e. China and to some degree in Mexico) and slow-growing but mature economies (the USA, Japan and the EU). In recent years, the freight transport sector has been less than well researched, and many books on transport economics and transport analysis tend to focus on passenger transport. With the exception of the great work 'The Transport Revolution (2010): Moving People and Freight by Without Oil' by Gilbert and Perl, Rodrigue's 'The Geography of Transport Systems' (2017) and Loo's (2018) 'Unsustainable Transport and Transition in China', many transport-related books tend to neglect the field, a fact reflected on book titles within the field. There are two reasons for this neglect. First

© Springer Nature Switzerland AG 2020

D. Bonilla, *Air Power and Freight*, SpringerBriefs in Energy,

https://doi.org/10.1007/978-3-030-27783-3_6

is the lack of primary data on the freight transport sector. Perhaps governments should pay more attention to publishing timely statistical material of the sector. For example the US official statistics on the field, surveys and websites tend to be rather outdated in comparison to Japan's or the Eurostat's statistical practices. Similarly, Eurostat still needs to offer greater detail of the industry since the agency is unable to report data on the sector with sufficient details. One reason for this is the confidentiality needs of the sector due to the highly competitive nature of supply chains and freight flows.

The second reason for the neglect of the field is its high level of disaggregation and interlinkages across industries and activities. A researcher can observe changes in freight transport by indirect means which are by no means perfect. One such way is by observing changes in the geographical spread of warehousing activity or aggregate data on new warehouses or through macroeconomic indicators of trade. The third reason is the neglect of the field in transport studies in Anglo-Saxon Universities and Transport Research Institutes which tend to focus on the passenger sector. This seems less pronounced in Latin America and Asia. In the economic history field, the freight sector is widely taught, but that neglects studying modern changes in the industry. The latest changes and innovations in the industry are commonly taught in Business Schools throughout the world, and it seems Schools of Economics or Geography fail to prioritize the subject. The Amazons and Alibaba's or Apples of today cannot exist without a dynamic and smooth freight transport network. To do so timely research on the sector is needed to help achieve a seamless freight transport system. More should be done by the economics and geography professions to teach and research the subject of freight transport.

Trade activity, correlated to export and import destinations, is a powerful source that will feed freight transport activity, inversely freight transport has trade creation effects. The combined effects of China's industrial capacity, its population and Europe's and the USA's household wealth mean that new freight transport demand will continue to grow in the three continents putting pressure on today's infrastructure capacity, on the environment, on urban air quality and on greenhouse emissions. Freight transport is a derived demand since goods need to be exchanged in the search for value or for meeting other demands.

Looking towards the future, it is likely that the rate of growth of freight transport will slow down for three reasons: mainly protectionist policies emanating from Washington, USA, which may curtail China–US trade volumes, the unforeseen effects of Brexit and the renegotiation of NAFTA now the USMCA. These geopolitical shifts are likely to cut trade among the major trading regions and weaken global supply chains and thus impact negatively on the sector as derived demand declines, and it is felt on the freight transport industry. These new trade policies are likely to change the destination of exports. On the positive side, the impact of e-commerce, robotization and the further uptake of containers will stimulate demand for road freight flows.

Chapter 2 provides an overview of key freight markets in the world and examines the road freight transport sector, its level of rebound effect for Japan, The U.K. and Denmark.

This is the only analysis that uses evidence from country regions from separate continents to study the trends and changes of the freight transport sector during 1950–2016. In the chapter we find that the USA shows the smallest increase in road freight in the examined economies; however, the USA ranks almost first in the volume of freight moved in our sample. Japan registers the largest decline in road freight transport, and China shows that largest increase of activity of that sector. The rate of growth of 1960s Japan is being mirrored by China in recent years. The modal split shows that China has a more balanced modal distribution for moving freight than the competing nations and that the balanced modal distribution of that nation is likely to have avoided an even larger increase of road freight flows. The modal split is largely dominated by road freight sector in the majority of the economies examined. The EU and Germany register growth just above the US and Mexico levels while the UK registers one of the smallest rates of increase in our sample.

We identify five key factors that determine freight transport activity: population density, trade, how far trucks travel and their average length of journey and the degree of competition among the transport modes.

Our examination of average length of haul finds that China and Mexico are both catching up with the levels of distance travelled by the US-based trucks calling on further comparative work on the complex relation between the vast territory (space) of these countries and their freight transport flows.

Trucks are travelling longer distances more than ever in most of the economies examined in the period considered. Worryingly we find that the distance travelled by the EU-, the US- and Mexico-based trucks are getting longer which produces even more congestion on highways and cities. China's freight transport sector is also registering longer distances for trucks which is likely to stimulate energy use and greater road congestion of the entire road transport sector. Japan is the only country in our sample that shows a decline in distance travelled by trucks, out of all economies.

The cross-country comparison shows that the per capita level of freight moved domestically by road in many countries is quickly catching up with the US level. The latter is the top performer in terms of freight moved per capita. The freight intensity of economies is highest in the USA historically; however, convergence among key industrial countries is occurring fast. The intensity is declining fast in Japan and the USA but remains stable in the EU. Mexico's freight intensity declines but not as much as the USA has done. The highest intensity can be seen in China while the UK level is also falling. It is unknown how much the rebound effect may have shifted the modal split and thus how much the rebound has favoured road freight transport. Further research is needed in this regard.

A number of conclusions can be drawn from assessing the empirical evidence on the rebound effect for the four countries. First, estimating the direct rebound effect provides evidence of the effectiveness, or lack thereof, of the energy efficiency of moving cargo by truck; however, the effect of energy efficiency on the volume of road freight flows needs further research. Second, the analysis also allows estimating the effect of fuel prices on the energy efficiency level of the cargo moved within each of the countries.

Third the analysis does capture a general pattern: the rebound effect declines over time. Our findings confirm work of Van-Dender (2007) and of Matos and Silva (2011) and of Stapleton et al. (2016). The latter study cites 9–36% of rebound effect values with a median of 16%.

Among the three economies, the analysis shows the rebound effect has been greatest in the UK. The gains in fuel economy of 1% lead to a decrease in fuel consumption of diesel by 0.7% for the UK. This is not a proportional decrease in fuel consumption because the rebound effect is 29% for this nation. In contrast to the UK, the estimates for Japan and Denmark on the rebound effect decline in the periods of low oil prices.

In addition to the analysis of the rebound effect, we have observed that both on road fuel economy (large trucks) and the carbon intensity of truck freight (large trucks) continue to fall in UK. China's and the USA's carbon intensity of road freight is also the lowest in the sample; this is because the vast territory of the last two countries requires long distance road freight services which requires fewer stop and start cycles helping reduce energy use. That reduction is actually an improvement in the mitigation effort of carbon emissions. Although standards on conventional pollutants do apply in these economies, truck fuel economy is not regulated directly. The total energy use of trucks in most economies will continue to rise because of four factors. First, the absence of fuel economy regulations (except China); second, a low average vehicle load which increases truck trips; third, a larger hauling distance which stimulates fuel use further; and fourth, the popularity of vans which have worse fuel economy performance (more litres per kilometre driven) than larger trucks do.

Future work should assess the rebound effect of freight transport at the level of commodity (primary products, foodstuffs, cement, etc.), but this requires detailed data on freight transport flows which is not normally publicly available. Notwithstanding these limitations we have produced estimates on how entire economies are likely to choose to buy freight transport services and on how the sector behaves in the realm of distance and time. Policy measures should encourage a transition to sustainable freight transport through shifting to alternative transport modes (which requires an integrated transport policy), the adoption of alternative vehicle technologies and the improved utilization of trucks, as well as the adoption of best practice programmes for vans.

In Chap. 3 we turn to the future of road freight transport in Europe by applying the backcasting method in order to attain sustainable road freight practices. Despite efforts, published in many White Papers by the EU, to reduce fossil energy use, cut both emissions and road congestion, the sector continues until today to rely on the internal combustion engine. However, the exercise presented in the chapter shows that backcasting work can guide policy makers. To guide them towards sustainable freight, practitioners should ask three questions. First, how far trucks ought to travel within cities and among these by 2050? Second, how much cargo volume can Europe's economy and infrastructure cope with? Third, how much GHG emissions can European societies believe is acceptable? The key premise of this chapter is the belief that transport policy makers and infrastructure planners should ask first what the socially desirable target is for the three dimensions just mentioned. These three targets need to be agreed within a consensus in order to take strategic decisions in transport planning today. Doing the latter requires the vision which we present here. A different growth path of road freight can emerge as a reaction to the effects of the megatrends. The effects of the latter and uncertainty in many fields calls for a vision for sustainable freight which builds a common understanding about the sector's future leading stakeholders to action.

The backcasting approach gathered views and beliefs of industry players, logistics firms and transport Ministries from the European Union. This was done for three reasons. The first is to identify the key megatrends. The second is to set targets to 2050 for the freight sector, and the third is to formulate actions and policy measures that are needed to close the gap between the vision and the current path which is unsustainable. Building the vision requires the strategic conversation technique allowing a variety of viewpoints of the stakeholders.

To achieve the vision targets, a variety of actions are recommended in 12 policy areas. Out of these policy areas, cutting freight traffic and reducing average distance hauled are key, and these targets can be achieved through four measures: congestion pricing, fuel taxes, the efficient use of vehicles and shifting the modal split favouring the railway mode. Further actions to minimize emissions are suggested in the chapter. In short, the measures recommended in the vision are related to the technical, managerial and supply chain domains. Other measures are based on economic and European transport policy.

The vision is an alternative to traditional forecasting techniques, and it depends on setting targets. Unlike the DHL Deutsche post (2012) and the US TRB (2013) studies, this analysis includes a wide array of stakeholders from more than ten key European economies and for many sectors of the Freight sector.

The vision proposed in this chapter provides legitimacy for the actions recommended by encouraging corporate and non-corporate partners to consider the limits to economic growth. Our work could be extended to assess further the role of uncertainty by including three or more explorative scenarios rather than including a single desirable future in the analysis of the freight sector to 2050.

In Chap. 4 we assess the air transport sector (freight) of the EU which complements the analysis of the Europe's road freight transport sector. This study illustrates the combined role of changes in the geography of airfreight, connectivity of airfreight flows, and sustainability of airfreight. Airfreight activity has become a key sector for time critical products, despite high jet fuel prices; the sector is also important for high value products. Airfreight has grown enormously, in turn, generating the need for more airports, whilst increasing the need for better use of hub and spoke networks. Corporate actors of airfreight have adapted their services and geography to compete for markets. These actors have modified the role and the position of airports. For example, urban airports do generate logistical activity which confirms the findings of Bel and Fageda (2005). These airports are being used as part of the niche strategies by some airlines; however, the airfreight sector has the ability to be the maker and the breaker of city economies repeating the process of road freight in shaping cities in the past.

Additionally we can say the high concentration of airfreight flows in a few of China's cities reflects a possible law of Zipf (1949); that is they follow a power law behaviour. Although further analysis is required in this regard.

As Kasarda and Lindsay (2011) have argued airfreight flows have the ability to alter the regional economies through its influence on the location of airports, on logistics and manufacturing firms. This is the situation in North-West Europe.

The above changes have multiple consequences for the future expansion of airfreight and for the environment-economy nexus. If current rates of growth in continue into the future, airfreight CO_2 emissions (the main GHG) will also grow since fuel switching is unlikely to occur. This means society will need to choose between the benefits given by the speedy deliver of goods and meeting environmental goals i.e. cutting GHG emissions.

While limits on emissions of GHG can cut airfreight flows, other megatrends will do the opposite: stimulate growth of airfreight in future. For example the rise of e-commerce combined with the need for speed can be strong stimulants for growth; the air transport is uniquely apt to benefit these two megatrends. These megatrends should form the basis to lead to actions by firms of the air cargo industry.

In Chap. 5 we focus on the air transport sector of the fast-growing economy of China which bears the potential to shape the global order of trade and thereby of the airfreight market. This potential is absent in the mature of economies of Europe and in the latter's freight transport industry. Road freight transport feeds the air transport and the notion of coordination has not been taken up in sufficient depth: future research will have to focus on these linkages. Although the practice of "air-trucking" is more widespread than assumed in and outside of China, it is not well researched.

Like cities do, airports are advancing on each other alongside air power and freight. This chapter examines airfreight transport networks from the optic of international trade and the spatial structure among China's cities. In the chapter, the airfreight network is superimposed on the urban network of China's cities bringing a new method to observe airfreight transport. The airfreight transport network (a) supports export growth of high value products, (b) untaps the geographical advantages of manufacturing establishments within China and (c) strengthens the existing system of dominant cities or centres in China. The air transport network that is emerging can potentially (a) lead to economic vitality within regions and (b) reduce or widen the regional inequality in incomes within China.

To examine what determines China's airfreight transport network within the period of 1975–2015, we closely examine the key factors such as the type of trade, air transport deregulation, population growth and density, urbanization rates, proximity to markets and competition from the perspective of the rail and road modes and e-commerce. Further factors as far as the ATNC include demand for timely and speedy delivery of goods.

Our evidence suggests the historical growth of airfreight flows of China has tracked closely that of domestic and international trade.

A key driver of air cargo trade has grown enormously in the last 15 years taking China from small player in international trade to a major exporter supported by the air mode. In short airfreight activity can be explained by the import and export activity as well as by the geographical spread of manufacturing establishments.

Overall five outcomes emerge out of the quantitative analysis on China's airfreight sector. One result that stands out is that airfreight flows and population density are positively and highly correlated in contrast to flows of surface freight transport. Further research is needed to explain the links among the population density of cities, airfreight markets and connectivity.

In addition to aggregate analysis of airfreight, we investigated the relationships among China's regions and air cargo traffic using a cross-sectional data set on airport cargo traffic, wages, population density, regional trade, high technology and time-dependent exports. The socioeconomic variables exhibit positive impact on air cargo in cities throughout China. Our results hold that a proportionate relationship between city size and air cargo holds as in *many* other studies. Average wages exhibit a positive relationship with air cargo. Airfreight demand and flows can also be explained by the connectivity levels for a great number of cities, but the positive relationship does not always hold for all the cities.

Our evidence reveals that China's air transport network is heavily unbalanced, reflecting the geographical distribution of manufacturing, high-tech and food production activities.

As China continues to expand its air transport infrastructure and widen its aircraft fleet, its cities will continue to expand. Cities will continue to have an impact on the air transport network worldwide; the reverse is also true: transport will continue to be the maker and breaker of cities (Clark, 1958). Within China the impact occurs via production and consumption of high-tech products. Trade relationships within and outside of China will determine the volume of airfreight and the shape of the ATNC. It is important to identify which cities are poorly, or well, connected to the ATNC, and this chapter has tried to do so. It remains to be seen how China's Silk Road will impact on regional trade with its neighbours to the East (Russia and India, Mongolia and others) and to the South (Vietnam or others) thus change forever China's air transport sector.

The book closes with a thought: Where will air transport and road freight flows end if trade has been part of human history for more than a millennia? The answer is that freight flows are likely to continue to expand, contributing to the global economy and further increasing GHG emissions.

References

Banister D (2018) Inequality in transport. Alexandrine Press, Oxfordshire, pp 1–240

Bel G, Fageda X (2005) Getting there fast: globalization, intercontinental flights and location of headquarters. http://128.118.178.162/eps/urb/papers/0511/0511008.pdf. Accessed January 2014

Clark C (1958) Transport the maker and breaker of cities. Town Plan Rev 28(4):237–250

DHL - Deutsche Post (2012) Logistics 2050: a scenario study. http://www.dhl.com/content/dam/
    Local_Images/g0/aboutus/SpecialInterest/Logistics2050/szenario_study_logistics_2050.pdf.
    Accessed June 2015
Gilbert R, Perl A (2012) The transport revolution: moving people and freight without oil. Earthscan,
    London
Kassarda J, Lane GLA (2011) Aerotropolis: the way we will live next. Penguin Books, London,
    pp 1–449
Loo BPY (2018) Unsustainable transport and transition in China. Routledge, New York, pp 1–226
Matos JF, Silva FJF (2011) The rebound effect on road freight transport: empirical evidence from
    Portugal. Energy Policy 39(5):2833–2844
Rodrigue JP (2017) The geography of transport systems. Routledge, New York
Stapleton L, Sorrell S, Schwanen T (2016) Estimating direct rebound effects for personal automo-
    tive travel in Great Britain. Energy Econ 54:313–325. ISSN 0140-9883
US TRB (Transportation Research board) (2013) NCHRP report 750: strategic issues facing
    transportation, Volume 1: scenario planning for freight transportation infrastructure invest-
    ment. Report prepared by Caplice, C., Pladnis, S., MIT. Research sponsored by the American
    Association of State Highways and transport officials in cooperation with the Federal Highway
    administration, pp 1–149
Van Dender K (2007) Fuel efficiency and motor vehicle travel: the declining rebound effect.
    Energy J 28(1):25–52
Zipf GK (1949) Human behaviour and the principle of least effort. Addisson–Wesley, Cambridge

Index

© Springer Nature Switzerland AG 2020
D. Bonilla, *Air Power and Freight*, SpringerBriefs in Energy,
https://doi.org/10.1007/978-3-030-27783-3

Printed in the United States
By Bookmasters